HOME OF THE ANGELS
&
NASA'S TETHER
EXPERIMENT

By
HENRY KROLL

authorHOUSE™

1663 LIBERTY DRIVE, SUITE 200
BLOOMINGTON, INDIANA 47403
(800) 839-8640
WWW.AUTHORHOUSE.COM

First published by AuthorHouse 12/22/05

IISBN: 1-4208-7895-6 (sc)

Printed in the United States of America
Bloomington, Indiana

This book is printed on acid-free paper.

This is a provocative chronicle about alien beings, human hybrids, secret government alliances with extraterrestrial entities, and the military complicity essential to promote secret agendas. The author examines overwhelming evidence of flying saucers, alien-human hybrids, alternate sources of energy, secret repositories of genetic and written knowledge, the space-time continuum, and the quest to uncover the star-gate; thought to be in Iraq. Fasten your seat belt; this is a voyage at warp speed.

PREFACE

President Ronald Reagan knew that destroying the Evil Empire was more important than fighting global warming and diverted the funds earmarked for the Mars Projects to building new weapons to destroy the Soviet Union. Before mankind can get off the planet he must create an environment conducive to benign human development. In order to do this one must eliminate threats to your very existence. The Soviets resisted the change by withdrawing their cooperation from Alternative 3 and so the world was left with no contingency plan to save humanity from cataclysmic events such as massive meteor strikes, nuclear war, or global warming.

The goal of the Alternative 3 and 4 projects were to establish manned bases on the Moon and Mars for the purpose of preserving a few people to re-seed the earth. To understand how they may have actually accomplished this feat without anyone knowing about it and doing it on a reduced budget.

After the President Reagan bankrupted the Soviets it left the nation three trillion dollars in debt. There was no money left for such lofty and noble projects as Alternative 3.

Earlier studies estimated that the planet's temperature would rise high enough to cause the melting of the ice caps by the year 2000. Cutting back air pollution in the 1970's did slow down the process to a small degree enough to gain about twenty extra years before disaster struck. To slow down further global warming a system of spraying aluminum powder over the large cities was devised to reflect solar radiation back into space. After the HAARP project went on line they pumped massive amounts of microwave radiation into localized areas of the ionosphere thereby heating it and causing it to rise which in turn caused the cold air of upper atmosphere to dip down cooling other areas.

No one had foreseen the damage to the ozone layer caused by unlimited use of fluorocarbons and release of carbon dioxide from burning carbon-based fuels. A massive hole opened up over both poles causing huge amounts of cosmic radiation to flood the planet further aggravating the problem. Something had to be done or the planet would bake from solar radiation. It was clear that some outside intelligence had declared Mars off limits for human colonization.

After we started exploding atomic bombs in the forties and fifties a wave of flying saucers came from everywhere. The electromagnetic pulse (EMP) of the blasts seemed to interfere with instruments aboard some of the saucers causing them to crash. Over the years our government collected several alien flying saucers and their occupants. They eventually stuck up a deal with the Aliens, which allowed them to continue with their agenda of the genetic manipulation of the human race and the alien-human hi-bread experiments. The US government agreed to furnish them with the breeding stock. After being born of surrogate mothers, the babies were adopted by families living on various military bases. The human hybrids attended school at various military bases and eventually joined the military. Some even fought for their country like everyone else. The only difference between them and us is that some of their internal organs are not the same.

The deal with the aliens went sour. Our government started to send them inferior humans and they stopped sharing their technology. This may be the reason President Reagan made his alien threat speech, "I wonder what would happen if we were threatened by...something from up there, a race of beings not of this earth."

Since the time of the Reagan Presidency we have had three presidents whose primary goal was to secure new sources of petroleum for the big oil companies who bankrolled their campaigns. The status quo freaks are still killing off inventors of new energy technology to protect their investment in big oil. The United States was running out of energy.

The safeguarding of human knowledge and the genetics of other life forms through the storage of frozen embryos stored on other planets is no longer an option due to economic reasons. The Secret Government however has constructed certain repositories of genetic and written knowledge here on earth, which are completely unknown to the private sector.

Unknown to the general public, The Secret Government has been aware for some time of ancient high technology given to man by the gods of ancient Sumeria. Still buried under the sands of Iraq may be the remains of an ancient star gate, a device that employs time travel technology, permitting individuals and equipment to pass into another dimension outside of the space-time continuum. Once you are in it is possible to travel across the galaxy, or even into another galaxy, instantaneously, and emerging back into space-time on another planet. Hence the real reason for the United States to invade Iraq.

According to Dr. Michael Salla, it probably is exactly what happened! Such a fantastic theory would not be expected from someone whose entire career has been spent in academia, pursuing conventional political and diplomatic studies. Dr. Salla is an Australian national who obtained his MA in Philosophy from the University of Queensland in 1993. After spending two years as an associate at the Center for Middle Eastern and Central Asian studies in Australia, he joined the faculty of the Department of Political Science at Australian National University in Canberra, as a lecturer. Arriving in the US in 1996 he was appointed at the School of International Service at American University in Washington, DC where he remained until 2001. Currently he is a Researcher in Residence in the Center for Global Peace at American University.

Dr. Salla published his star gate theory on his web site along with several other papers, which exhibit first-rate scholarship, elaborating on worldviews and events. Salla thinks that the extraterrestrial presence should be a major factor in all political decisions and actions. One of his 10,000 word research papers is titled, The Need for Exopolitics: Implications of Extraterrestrial Conspiracy Theories for

Policy Makers & Global Peace.

According to the Intruder Perspective, the aliens are here for their own purposes, and their abduction and hybrid program activities are intrusive and callous to human concerns. Although they pose no direct military threat, this attitude would tend to justify defense concerns and the development of weapons with which to confront them.
Taking the manipulator perspective, a case can be built that the aliens have been manipulating humans covertly since we first appeared here. There is evidence that this has been conducted through their human proxies via secret societies and support of a human controlling elite class."

From the helper perspective, it appears that the aliens are here to help us to grow in consciousness and solve our political problems. Much evidence supports the view that the ET's play an important role in encouraging humanity to achieve peaceful resolution of international conflict, and in preventing nuclear proliferation and the use of other destructive weapons."

According to the Watcher Perspective, the aliens are agents of a larger galactic organization and are here to observe a great experiment that may help other planetary societies if we resolve it correctly. There aliens are all-powerful, and although non-interventionist, are ready to step in if absolutely necessary. Of these four possibilities, he claims that the strongest evidence supports the Intruder and the Helper Perspectives, and therefore government Policies should actively address both of these.

First we must create the field of exopolitics so that schools and colleges may prepare diplomats for dealing with the extraterrestrial presence, and also the ET's themselves. This would require highly specialized training. Second, and perhaps most important, he argues for the pressing need for immediate disclosure so that exopolitical leaders can bring the best academic minds all over the world into the problem-solving matrix, rather than must those now secretly employed by the government. He says, "This would lead to a more representative decision making process in contrast to what the evidence suggests is a restrictive decision-making process on the ET presence run by a small number of government officials appointed in a manner which raises serious concerns over their accountability, constitutional status and lack of congressional oversight."

The Military Industrial Complex probably does have ways of identifying and utilizing brilliant and talented individuals, as in *A Beautiful Mind*. Dr. Salla claims that they have brain- enhancement, technology that can raise an IQ by as much as 50 points. But, there are obvious disadvantages to operating in a climate of secrecy, fear and intimidation, and we are all being shortchanged by this policy."

Only a free and open discussion can elicit the responses appropriate to the incredibly complex problems and fabulous opportunities presented by the existence of several alien races on the planet, displaying powers and technologies that are thousands of years ahead of us. He believes that waiting for official disclosure will cause us to loose valuable initiatives and opportunities for change, and could actually be disastrous. The governments of the world should reveal exactly what weaponry and defense measures they have developed for dealing with the alien menace."

From time to time NASA space shuttle cameras catch a glimpse of three-mile-diameter ships coming in and out of higher dimensions. At first it was thought these were frozen globs of water however they had form and shape such as one or more square notches cut out of one side. They also could disappear and reappear from higher dimensions at will. For a time scientists argued that they were some kind of lens reflection in the camera or reflection from the sun however their sheer numbers, form, and shape and their ability to move about negated all of these arguments. They all have a black hole in the middle with a blue spiral of electricity pulsing in and out of it.

Since World War II our government became increasingly aware that the earth is a hive of intelligent beings living underground and in the upper atmosphere where they exist in some sort of higher dimension or energy field. These extremely advanced life forms seem to have always been there, having the capability of manufacturing and providing for all their needs by literally creating them out of thin air. They use the technology of the monatomic atoms, which I describe in my book, Gold Of The Gods.

Gilgamesh was the hero of an ancient Sumerian epic poem, The Song of Gilgamesh, written about 2900 BC. He was the king of Urik, a city in southern Mesopotamia, who sought immortality by searching for the Stairway to Heaven and the abode of the Gods. Ultimately, he ascended a ladder and went through a gate (star gate), and entered a New World where he met the long-dead Utnapishtim (Noah) and questioned him about the gods.

Dr. Salla believes that the star gate will be found at Uruk, where German archaeologists have been digging for many years at the invitation of Saddam Hussein, and have unearthed an ancient city. He thinks that the star gate may have already been unearthed, but cannot be used because it probably takes about five years of consciousness expansion training before someone can walk through the gate. This is consistent with the modus operandi of the now famous Montauk experiments in time travel, which were driven by the enhanced mental power of one man, Duncan Cameron.

If we are being embroiled in a weapons escalation race, fueled by competing ET factions fighting for domination of the planet, and are being used as pawns in this struggle, then it is clearly time to make the whole thing public and allow the people to have a voice in preventing their own doom in an apocalyptic war, i.e., Armageddon."

CHAPTER ONE

Thousands perhaps millions of super intelligent beings live aboard these giant ships as picture in NASA footage taken from the space shuttle. The size of these ships is from two to three miles in diameter. One can easily estimate their size by comparing the diameter with the twelve-mile-long cable attached to NASA's long-wire-tether satellite. During the 1996 tether experiment hundreds of these giant ships were swarming around NASA's tether cable and satellite.

The ships are invisible to the naked eye because they exist in a higher frequency or energy state than our optical receptors are capable of seeing. In other words our eyes don't see into the higher ultraviolet light frequencies. NASA cameras however are wide spectrum cameras with special receiver chips capable of looking into the higher light frequencies. I believe the ships exist in an even higher dimension than NASA's cameras can see and only became visible during the tether experiment for observation purposes. They exist in the higher energy states for the purpose of reducing their mass so that they do not attract to gravity and it also makes them impervious to meteorites. Meteorites can pass right through their higher dimensional state.

Note the blue spiral emanating from the black hole in the center. The blue spiral of electricity pulses in and out of the black hole once every two seconds. All the ships have one or two square notches cut out of the outer edge. These are probably huge access ports and may have something to do with aerodynamic flow to keep the craft oriented in a given direction when entering the atmosphere but I wouldn't bet on it.

The people living aboard are so far advanced that they can provide for all their needs using the high voltage in the ionosphere. I believe they can manifest anything their hearts desire much like the food replicators aboard the 'fictional' Enterprises in the Star Trek television series.

In 1996 NASA decided to send up a satellite to measure the voltage in the upper atmosphere about 75 miles above the earth and then film the experiment from the space shuttle. NASA's long-wire tether experiment measured the voltage 75 miles above the earth at 12 to 14 million volts---so high in fact that it burned the tether cable in half in a couple hours.

Living in the higher dimensions seems a natural evolutionary goal for humanity given the various cataclysmic events that affect earth from time to time. Such events that might periodically wipe out life on earth are nuclear war, pole shifts, meteor impacts, solar flares, and close encounters of large gravitational objects that cause massive tidal waves thousands of feet high. Living in the upper atmosphere in a higher dimension would allow living beings like us to escape such disasters.

The beings living aboard the ships are truly the Adoni, the Malakeim, the Nephilim, the Sarafim, and the Metatron, Angels of old. Since they are so far advanced like any super-intelligent race, which has solved most of it's political problems they probably become bored and from time to time so every thousand years or so decide to meddle in affairs here on earth. I believe there is some Communication through prayer and they probably do respond from time to time to help us out of a difficult situation. To them we are like pawns in a game whose outcome has already been decided. If things are not going their way they could easily kill off a few million of us. There is no way we could retaliate because they exist in a higher dimension.

Such inter-dimensional ships as these could exist around thousands of other planets in other solar systems throughout the galaxy. There are billions of other galaxies so it would be interesting to find out just how extensive this civilization is.

Obviously they need us or they would have exterminated us long ago. It seems we enrich their lives by vicariously experiencing our emotions. Much like we watch violence, dramatic action and drama on television they look in our daily lives to experience our trials and tribulations and laugh at our stupidity. To them we are like watching the Fox channel.

Apparently, we are a binary simbioent. We are they and they are we, only in a higher energy state. We are obviously symbiotic with them as evidenced by our need to believe in a higher power (religion) and our need for prayer.

Such a life is not without problems. Any society that has evolved to the point where it can read minds and receive messages from the collective is restrictive and stagnating.

One has to constantly guard his or her wrong thoughts. Wrong thinking might be punishable by expulsion to the earth's surface or worse yet, Hell located beneath the earth's crust. Thousands of years of controlling one's thoughts and emotions might lead to loosing one's emotions entirely. They might acquire a sort of bland mental state, not caring and having few emotions. Over time for lack of stimulation they might become cruel even sadistic.

Human emotions can be beautiful but they also contain the potential for violence and destruction. Obviously the beings living aboard the higher dimensional ships see the need for human emotions and have elected to let us live here on earth with little or no interference or supervision. They gave us dominion over all the animals and plants and we are screwing up the world again. It seems that they enjoy our emotions vicariously and in return they have assumed the role as our protectors defending us from other alien interference. Whenever some outside alien race decides to interfere with our lives they step forward out of the higher dimensions to demonstrate the awesome powers that they possess; however I don't believe they do this until we ask them for assistance.

HOW THEY STAY UP THERE

Townsend Brown was flying circular convex-shaped craft tethered to a pole. When the military saw the discs flying around in a fifty-foot diameter circle they made the project top secret. It is proven the charged convex shape ionizes the air in front of the craft pulling it through the medium. In the case of the large flying discs with the blue spiral of electricity pulsing in and out of the center column there is more than high voltage involved.

According to witnesses who claim to have been inside a flying saucer there is a circular, central-column that is believed to be some kind of resonant transformer coil. It also serves to support the center of the vehicle. Some reports say there is a sort of turbine generator mounted in the center. Witnesses who have discussed this power plant with extraterrestrials themselves report that the system is used as stationary power plants for electrical energy on their home planets. This may be true but I do not think the turbine has anything to do with lessening the affect of gravity on the vehicle.

This is how I think it works: The central column is wrapped with a caduceus coil (a coil with two windings wound in opposite directions from each other). When the coil is energized it creates a bucking magnetic field pumps out dark energy. This dark magnetic energy is pumped over the outside skin of the craft at the rate of two cycles per second up to a hundred cycles per second when going into hyper-space. Seventy three percent of the Universe is composed of dark energy. This dark energy crates a dimensional field around the craft. All this takes very little power because it is controlled by frequency. You could say the system is frequency modulated (FM) to maintain optimum affect.

The skin of the ship itself is covered with a diode like material made of bismuth coated aluminum. Bismuth is an excellent radiation shield. The surface repels space radiation and melts momentarily when impacted by a micro- meteorite. In the sub-zero of space the bismuth metal which is momentarily liquefied quickly closes the hole.

The reason witnesses cannot find any kind of propulsion devices or generators on board these ships is because there isn't any! The power is supplied by a small storage battery, which is recharged by various means including solar cells sensitive enough to take in the dark energy from deep space. Tesla and others have tapped into this energy in the past.

Inside the craft witnesses' report that there are windows evenly spaced around the outside wall of the craft about elbow level. The windows are about one foot thick and have an iris type of shutter so that when it is closed, it allows electrostatic charge to flow evenly across the outside of the craft.

There are 'electro-optical' lenses, (TV cameras) arranged around the outside of the craft pointing in all directions. TV like monitors are placed on a console where the pilot can observe all areas at the same time. Also, they are equipped with magnification lenses, which are used without changing positions.

MIND CONTROL SOCIETIES

I have no doubt that at one time humans had the ability to communicate telepathically. I also believe that we evolved a mind-control society here on earth. This is the reason why we have such large brains. Such a society seems to be a natural evolutionary trend in human development. Our pets still use this form of communication by staring at us until we do their bidding. Examples of such wants and commands are to give them food or let them outside.

For some reason such as a cataclysmic event or a great rebellion against the ancient mind-control, societies most of us were exterminated. Before the Biblical patriarch, Moses time mankind was enslaved by telepathy. We rebelled against this enslavement. This is evident in Moses leading his people out of Egypt in the book of Exodus. Such a society is slavery because one has to watch what one thinks at all times or is punished for wrong thinking. The flood further killed off what was left after the great rebellions. Mankind then degenerated back to living in caves and hunting with bows, arrows and spears for a time.

A thousand years of living in a mind-control society would produce people with little of no emotion because you have to constantly guard your thoughts. People with little or no emotion might become heartless and more willing accomplices of the system. Cruelty would abound toward any dissenters and much like the Catholic Inquisition, torture, and murder would be commonplace. In other words it could be a living hell or it could be Heaven depending on how you look at it. Maybe this is the reason why only the pure of heart are accepted into Heaven and the wrong thinkers are condemned to Hell here on earth. Our government wants to enslave us completely today but they haven't figured out how to do it yet. See my book, The Frog is Cooked, subtitled, Why You Have To Work Two Jobs.

I believe the ARK is a remnant of an ancient mind-control society. A small group of Egyptians and survivors from Atlantis resurrected this technology after the flood and ruled the world with it another two thousand years. Fortunately for all of us Moses took it out of the great pyramid and hid it away another two thousand years.

In my book, Philosopher's Stone, I wrote that that the ark was used to carry the m-state white-powder-of-gold that is a room temperature superconductor. The ark itself is a metal lined wood box especially designed to carry a superconductor. The metal is used to isolate the m-state from outside magnetic fields. It is a difficult material to contain because it will even tunnel through glass. It has to be placed inside metal shielding to keep it from being energized by non-local magnetic disturbances.

Moses and other adepts who ate this material in their bread for many years had increased brain functions because it superconducts within the microtubules in the brain. Using only their brains at a distance they could create an electric discharge between the outstretched wings of the Cherubim mounted on the top of the lid of the ark. If you can amplify your brain waves a million times by turning them into an electric discharge using the room-temperature super-conducting m-state material then you can broadcast thoughts into the minds of others!

The ark was used for thousands of years as a mind control device and may not even have been invented in this planet!

The One World Government now wants to reinstate the mind-control society; only this time they plan to use crude electronic implants and the HAARP transmitter. Such implants could contain a microgram of explosive.

When they decide to terminate you all they have to do is push a button and boom your head disappears. You had better watch what you are thinking or else. The big problem with mind control societies is that eventually they destroy themselves because they stagnate. Nothing new is ever invented because the bureaucracy won't allow it. Bureaucracies like to keep everything the same as they are with no possibility for advancement. A thousand years of status quo is more than any society can endure.

According to Edgar Casey there may have been two different mind-control societies existing 25,000 and 50,000 years ago that destroyed themselves. What was left after the great ancient rebellions degenerated back to living in caves and again hunting with bows arrows and spears. Now we are evolving toward another such experiment. How many more times must we go through this cycle? When will we be smart enough to see what is going on?

Its time we became a type one civilization again, transcend ignorance, repression, greed, envy and lust for power, and get on with the business of colonizing other planets for the purpose of spreading intelligent life and benevolent life forms throughout the Universe. The status quo freaks are the enemy. They would kill everybody off to keep things as they are. This is contrary to what it means to be human. In order to become more human one need to grow, explore, and learn new things.

If we are to ever understand the people living on board the giant ships existing only 70 miles above us here on earth we have to start thinking in multiple dimensions. The first step to doing this is to realize that higher dimensions exist. The second step would be to increase our intelligence about a hundred fold so that we could communicate with them telepathically. By using monatomic iridium in the ORME state (mfkzt as described in my book, Gold Of The Gods) I believe it is possible. Mfkzt is used by your body cells during cell division to correcting your DNA to that of a twenty-year-old. It is a room temperature superconductor, which enables your brain to communicate with the Gods on the Plane of Shar-On. Ingestion of the high ward firestone (monatomic platinum group) increases the cognitive and perceptual abilities by stimulating the pineal gland to produce melatonin thereby assisting in the opening of the third eye. The pineal gland is also known as the "Eye of Wisdom."

The endocrine glands are named from the Greek verb "to arouse," are ductless glands, which secrete directly into the blood stream. "The Pineal Eye has been found to contain very fine granular particles, rather like the crystals in a receiving set." Is it some kind of thought receiver?

A hundred years of using this 'advanced' technology along with multiple demonstrations of nonviolence and they might allow us to meet them.

We have come a long way since the dark ages of the inquisition but we may still be on the wrong evolutionary path. The One World Government people want us to go back to the old system of Heritage, rule, rank, and privilege of the middle ages. It didn't work then and it won't work now. Mankind has no chance of advancement under such a totalitarian system of government. America was the world's last great hope to crawl out of the slime of a stagnating, Fascist, bureaucracy.

One thing is for certain about the higher dimensional ships is; they were extremely curious about what NASA was doing up there with the 12-mile-long cable stretched out 75 miles above the earth. Hundreds of them came down from the high-energy state that they normally exist in to become visible to NASA's high-resolution cameras aboard the space shuttle. My hunch is they were concerned that NASA's experiment might pose a threat to them or the knowledge they would gain would pose a threat in the future. Whenever a branch of our government gets involved there is always the possibility the knowledge gained will be used for corrupt or devious purposes.

Higher dimensions are easy enough to attain. When you completely fill an atom's electron orbits and add one more you kick it up to the next dimension disappearing from the third dimension. The ORME particles are easily pushed up to higher dimensions because they have so many electrons (double the number of a normal atom), and are easily sped up beyond light speed by a dimensional frequency field. ORME atoms extracted from dirt using the alkaline solution will have all kinds of regular atoms along with the ORME particles. Each one is a tiny time machine existing in it's own separate dimension. When you think about them they respond by giving off electrons and the regular atoms around them are kicked up in energy thereby transmuting them to the ORME state where they work together as one creating a dimensional bubble.

Anything that is within this dimensional bubble does not attract to gravity and is included in the dimensional time field.

In SPACESHIPS OF THE GODS I wrote that IRON is easily pumped up to the higher dimensions and that it is responsible for at least half of the earth's mass and gravity. Iron might even be responsible for more than half of earth's gravity due to it unique dimensional properties. The earth itself may have an iron core, or at least that is what some geologists' suggest. I personally think it has an iron shell at the core. Throughout the earth's crust there are minute iron specula and deposits of iron ore that add to the total iron mass and could be used to evenly distribute a higher dimensional field.

I believe it is possible to put the entire earth in a higher dimensional, mass-less state that does not attract to gravity and push it through space faster than the speed of light to another star system. This may have been done in the past because it appears that the age of both the moon and the earth is older than our sun but our scientists refuse to look at the data. They can't imagine how such a thing could be possible.

The blue spiral on the surface of the God Ships, (what I call them) is the key to maintaining a higher dimension. It is also used to create a directional gravitational field capable of accelerating the craft to unbelievable speeds. The Diameter of Earth is approximately 13,000 miles. The speed of light is 186,300 miles per second. It takes approximately .06 seconds for light or electricity to travel through the earth one way. To make a complete cycle the time is twice that or .12 seconds. Any multiple of this frequency can be used for antigravity. No one understands this except Tesla because all radios and test equipment are grounded to the earth.

CHAPTER TWO
THE DROPA STONES

The "dropa stones" are hundreds of stone discs found in western China. They all have a hole in the center and spiral decoration etched on it. They also have a square notch cut out of one side. In other words the description is identical to the giant dimensional ships that inhabit the upper atmosphere.

The Dropa stones were buried over the top of tiny skeletons 3 to 4 feet tall which are believed to come from a large flying saucer that crashed near the Chinese Tibetan border about 30,000 years ago. The stones were carved as reminders and grave markers. This saucer is believed to have come from the Sires constellation. Further examination of these stones has revealed micro text engravings and these have recently been translated. I hope to get an update on this before this book is published.

Native tribes living in the area are small in stature with large heads and eyes. Could they be related to the people from the stars? Legends of water Gods who traveled in water-filled craft abound in the area.

There is a cave cut into the rock in the vicinity of the dropa stones and also ancient iron pipes driven into the ground. Who created the iron pipes?

DOGONS

When I first read about the Dogons in National Geographic in the 1960's I completely discounted the Dogon theory as something that they made up to get attention.

The Dogon Tribe of western Africa claim to be descended from the Sirius star constellations and their data corresponds roughly to the dates of the water-filled craft that crashed on the Chinese Tibetan boarder." The Dogon shamans or priests reported that the Sirius star system is actually composed of three stars and their associated planets. Dogon priests reported that two smaller stars orbit around a central star in fifty-year-orbits.

In 1970 astronomers discovered that the star Sirius was composed of two stars. In 1976 astronomers discovered a companion star, which had a fifty-year orbit. It was not until 1980 that a third star Sirius C was detected and as predicted by the Dogons this star too had close to a fifty-year orbit.

THE GODS WHO CAME FROM SIRIUS

Ea. Is reported by the Egyptians to have come from the Sirius star system Sirius in a metal-hulled ship filled with water. The Dogon priests also describe the Sirians as traveling in a water-filled spacecraft. Dogon priests predict that a new star will appear in the sky, The Star of The Tenth Moon. Dogon legend report that they migrated from Egypt. Is this where they got their tremendous knowledge of astronomy?

If you were coming from the Sirius star system eight light years away you would have to bring along a large supply of fresh water.

Seven thousand years ago Babylonian mythology tells of a God named EA. He was a water god depicted with the body of a man and a tail of a fish. The Greeks knew him as Oannes. Oannes taught the Greeks mathematics, architecture, and astronomy. Greek legend said that he came up out of the sea and that he was a master of water. Did the Sirians cause the flood?

The Egyptian god Osirus is reported to have come from the star system Sirius. The great Egyptian God Isis, symbolized the All Seeing Eye, who is currently printed on your dollar bill, was the wife of Osirus. She too is reported to have come from Sirius. Enclosed in my book Gold Of the Gods is a picture of her floating on a cushion filled with the MANNA. See also the statue with the outstretched wings and the antigravity device strapped to her head.

Anubis was the half-dog-half-man who was the pilot who flew the ship from Sirius. The Sphinx may have a body of a dog and not that of a lion after all. By the way, Sirius is also known as the Dog Star!

It seems that the Gods had the ability to clone any life form they so desired. This opens up the possibility that griffins, gargoyles, phoenixes, unicorns, satyrs and mermaids really existed. Wouldn't it be nice if you could get your dog to wait on you instead of the other way around? Hey Fido, get me another beer out of the fridge will ya!

DIFFERENT RACES OF PEOPLE ON EARTH MAY HAVE COME FROM DIFFERENT STAR SYSTEMS

Current evolutionary theory doesn't hold water. True, there are primitive fossils dating back over three million years that resemble man but they don't come close to Cro-magnon or modern man. There is no fossil record of modern man. You would think that since he came at a much later sate there would be fossil bones that show this evolutionary change. There is however, some evidence that the European races may have absorbed Neanderthal man. It's as if modern man came from Mars.

Given the above information, it is possible that different races may have come from planets in other star systems. This means that the earth could be a melting pot of alien races from all over the cosmos. We may be the aliens and don't know it. The real indigenous races of earth may be the Bigfoot, Sasquatch and Yetti although there doesn't seem to be an evolutionary record of their existence here either. It is beginning to look like the Earth really is the Planet of The Apes.

The Terraga, Balanese, and thousands of different tribes on earth have ancient legends of coming to earth in a ship of some kind. The houses on the various Pacific islands are constructed as models of a star ship. The religion and culture of the people of the islands describe life as a journey into the physical world. One starts out in Heaven as a spiritual entity and journeys down into the physical for a brief stay then slides back up again.

Its time we became a type one civilization, transcend ignorance, repression, greed, envy and lust for power and get on with the business of colonizing other planets for the purpose of spreading intelligent life and benevolent life forms throughout the Universe. The status quo freaks are the enemy. They would kill everybody off to keep things as they are. This is contrary to what it means to be human. In order to become more human one has to grow, explore, evolve and learn new things.

I believe our goal should be to reach for the stas using dimensional technology. To utilize this science we may have to expand our conscious through the use of the ancient Hebrew MFKZT powder known by other names as m-state or ORME particles. This is the only know room temperature super conductor. We use it for cell division and brain sanapse. If we were to eat more of it we would become more intelligent. In my book Philosopher's Stone I list seven recipes which show how to make it in your kitchen.

FATHER CHARLES MOOR

The following is from my notes about Father Charles Moor, a Catholic priest who was on the Joe Segal radio talk show November 24, 2000. A four hour transcript of this show can be ordered by calling: 1-800-917-4278 and order tape number: 001124C. The price is $27.50.

"The Catholic Church now acknowledges the existence of extraterrestrials and is terrified of the implications. The Vatican now agrees that Zechariah Sitchin translation of ancient Sumerian texts is correct and is terrified of what the revelation of the Anunnaki Gods presence on earth might do to the Catholic Religion."

According to Sumerian texts the Anunnaki visit earth every 3600 years. The reason they came here last time was to mine gold and burn it into the antigravity white-powder-of-gold, the mfkzt. They tried to take it out of seawater but the process took too long and the supply wasn't adequate so they went to South Africa mine gold. There they created mankind to assist them. This is one of many versions of the story. The location may have been the present day country of Iraq.

MORE ET EVIDENCE

MAGIC SCIENCE OF THE FUTURE is a book by Joseph F. Goodavage published 1974 by Signet Classics, Mentor, Plume and Meridian Books. Mr. Goodavage has written and published dozens of articles, both for scientific and general audience publication. Formerly science reporter and writer for the New York Daily News syndicate, The Chicago Tribune and the New York Newspaper Guild.

"It may be that Homo sapiens are not all that unique in the cosmos. Perhaps there are great civilizations spread across the Galaxy (and beyond) composed of undreamed-of variations of the genus homo. While we discover more about the human family and the riddle of evolution and as the mystery surrounding man's true origins are increasingly complicated, the clearer it becomes that other species of (quite possibly) non human intelligent, rational beings have evolved equally complex brains and languages. There are undoubtedly great alien cultures of non-human civilizations somewhere in space. If this as likely as it seems then it's even more likely that representatives of those civilizations could have visited Earth at any time (perhaps during our Prehistoric periods) and left artifacts or some other trace of their presence. This theory is now regarded as scientifically plausible."

"Technology stands at the crudest, most cumbersome era of space travel. Our mighty cannibalistic rocket, fueled by million of tons of costly chemicals, are the "gas bags" of space voyaging. Compared to the fantastic vessels of the futures, today's powerful boosters hot-air-inflated linen bag, or to the Write Brother's stick-and-cloth prototype of the airplane."

"Ever since American astronauts began to leave traces of their presence on the moon and Soviet and American space probes started parachuting sensors into the atmospheres of Venus, Mars, and Jupiter it has become easier for us to understand how the same events could and quite possibly did happen eons ago."

"It began to look suspiciously as though we're intimately involved in a Grand Design which we're only dimly aware of. The pieces are beginning to fit however, and the clearer it seems that we are destined to explore and colonize the planets and then the stars.

As stupendous as it seems, such an adventure may be as common an event in each Galaxy as are hundreds of thousands of graduating classes on all educational levels throughout the world. If the schedule is kept, we will witness the withering and gradual elimination of competition and war as tools of international diplomacy, and the appreciation of man as a unique entity in the Galaxy.

"Aside from conquering, disease and increasing out knowledge, power, and the ability to bring human reproduction into balance with earth's resources, the major nations of our planet are desperately engaged in hammering out treaties to avoid nuclear holocaust that might forever extinguish all trace of man in the universe. We posses this capability plus the freedom to choose either course. Either we accept fully the responsibilities of stewardship of our fragile, delicately balanced planet or we do not.

"Are we prepared for the final examination before graduation from the planetary grammar school?

If we are to make it, it's imperative that we gain total control over the animalistic promptings of the right brain and stifle its negative biological emotions, fear, greed, and superstitions and begin to live in a spirit of mutual trust, cooperation, and love."

THE EARTH IS A HIVE

The golden-balls or plasma entities that are seen swarming around crop circles as they are being created exist as ordered plasma and live in the Earth's interior. Ancient maps of the area where crop circles have been seen in England have place names such as 'Golden Ball Hill'. Obviously these living and intelligent entities have been around a very, very long time.

Plasma is one of the most abundant mediums in the Universe. The interior of stars and most planets is composed of vast oceans of plasma. Is it so difficult to understand that living plasma organisms may have evolved over billions of years as organized balls of electrons and other subatomic particles? When you stop to think about it, our human body is an organized bundle of atomic and sub-atomic particles.

Also living underground in giant cities are the ancient humans, which are hoarding the ancient Atlantean, Asvin and Vimanna technology. They took a few ancient-flying ships with them when they went underground. They emerge from time to time from the underground bases located in the mountains of Tibet and near Mt. Shasta in northern California to inspect the damage the current human inhabitants have done to the earth.

Then there are the three-mile-diameter higher-dimensional ships photographed by NASA's high ultraviolet cameras from the Space Shuttle. These beings live in the ionosphere in a higher dimension in what religious groups would term Heaven. They utilize the 12 to 14 million volts of electricity in the upper ionosphere as a source of free energy. HAARP technology might pose a threat to their existence when it is used to heat large areas of the ionosphere for climate control.

Recently released NASA photos of Venus and Mars show buildings, bridges, and other structures. Apparently mankind had colonized all the inhabitable planets in the solar system at one time but a Great War between

Heaven and Earth and other natural calamities wiped out civilization. This has happened several times in the past.

CHAPTER THREE
THERE ARE HIGHER DIMENSIONS

MY SON HENRY

The following is a hint that higher dimensions do contain higher energy. My son who is in his thirties repairs computers for a living. He figured out how to substitute and cross-reference many parts to use for repairs. He buys many computer monitors surplus from the government for five dollars apiece.

One day he decided to build a Tesla-like, high-voltage, device. He connected the high voltage output of a cathode-ray tube (picture tube) to a homemade capacitor. I believe it was a bank of two-liter plastic pop bottles filled with salt water and wrapped with aluminum foil. He also made a two-foot diameter air-wound, caduceus-coil from a design he found on the Internet. The plastic bottles were grounded to a large sheet of aluminum foil that covered the top of his steel desk. This desk was grounded to his thirty-foot, aluminum-covered. 1950's trailer that he used as his workshop. The desk had a rubber-insulated top like so many of the military-style desks.

When Henry turned it on and went outside his entire trailer was enveloped in a blue corona. It wasn't his intention to put his entire trailer into a higher dimension but that is exactly what happened. He could feel the electricity in the air when he entered inside. While sitting inside the trailer he accidentally discovered that he could move the cursor of his computer screen merely by thinking about where it should go. His brother Bill and several of his friends tried it and they stayed up all night experimenting with various kinds of thought experiments. While inside the trailer they were able to move small objects across the room with thought or make them hover in midair. Whatever they thought about came true. Henry estimated the voltage output of his device was about 3.5 million volts. It had high frequency but not much current. It was enough current however to increase the energy state of the atoms of his trailer and the occupants inside. He had accidentally created a dimensional portal or field.

The next day Henry tried a weather modification experiment. Anchorage had the strongest winds they ever experienced. The wind ripped off roofs and blew down buildings. After the damage was reported on the news Henry got scarred and dismantled his device and never tried it again. Henry is a brilliant and funny guy. Now he is afraid of experiments that have an unknown outcome. He doesn't want to create a situation that might hurt people, which is good and he realized that delving into higher dimensions is too risky. He did however make a three million-volt wishing stick or wand.

His experiments prove that higher dimensions contain higher energy and link thought with the physical universe which can making things happen. We can do the same thing from this dimension but it takes a lot more thinking and you have to keep at it for a very long time.

The reason Henry and his friends were able to move

objects with their thoughts and make the cursor move on the computer monitors is because Henry had accidentally put his entire trailer into or on the brink of the next higher dimension. The high voltage and high frequency physically energized all those inside. When you are physically close to or completely in a dimensional bubble your body has a higher energy state. Your thoughts are amplified about a hundred times and can be used to control three-dimensional reality because you command so much more energy.

Henry and I are convinced we live on, "The planet of The Apes." Over the last fifty years the carbon dioxide content of earth's atmosphere has doubled. The government says there is no problem but they secretly release virus, anthrax, aids and other diseases to kill off humanity instead of fixing the problem. Some people say it is the hostile, Reptilian aliens that want to enslave us that are using wars, small pox vaccinations and countless other devious methods to reduce the population so they can enslave us.

What the government fails to realize is that the CO_2 is raising the acid content of the oceans. All the choral reefs around the world have stopped growing twenty years ago. Currently half of all the plankton in the oceans is gone. This means that more than half the fish are gone. About half the rain forests on earth are cut down. We are loosing an area of forest the size of Great Britain annually. This means that a good portion of the oxygen-producing organisms on earth is gone so the amount of carbon dioxide accumulation is increasing with nothing to stop it. Soon we won't be able to breathe the air and it will be too late to do anything.

Carbon dioxide is also an insulating greenhouse gas, which keeps heat from escaping into space thus contributing to global warming. The sea levels are rising destroying wetlands, which are nesting for all kinds of life including migratory birds.

My son says, "Once the plankton is gone the oceans will heat up because the solar energy that is being converted to food by the plankton won't be used up. Instead the suns rays will heat the water. Once this happens the oceans will evaporate at a higher rate, which will increase the amount of water in the air. Rainfall will increase in certain areas. At the higher elevations and latitudes snow accumulation will increase unbelievably plunging us into another ice age in about twenty years. Whatever the outcome, it is clear to me that we have ventured into uncharted territory!

TESLA

Antigravity devices have been invented in the past. There are many ancient Greek and Egyptian legends of people who could fly. On the cover of my book, Gold Of The Gods is a picture of Isis wearing an antigravity device on her head with wings taped to her arms to guide her flight. Ancient Indian epics speak casually of flying to and from various places and planets. From time to time ancient Asvin and Atlantean ships come up out of the ground near Mt. Shasta, California and northern Tibet to inspect what mankind has done to the planet.

Nikola Tesla was a present day genius that actually overcame gravity electronically and built a personal flying machine with which he could fly around New York City at night. Nicola Tesla was ahead of everybody, especially in his concepts of how matter was put together. I personally believe that our pursuit of Einsteinian physics actually held back our understanding of the nature of the Universe at least fifty years. The Nazi's burned Einstein's books and papers and kicked him out of Germany because he spoke frequently against Nationalism and supported Zionism. This is the reason the US adopted him. Einstein's E=mc2, theory of relativity and quantum theories do not take into account the subatomic forces and the neutrinos or the black either particles that make up ninety percent of the mass of the Universe.

In 1937 Tesla announced that he had a new formula showing that, "The kinetic and potential energy of a body is the result of motion and determined by the product of its mass and the square of its velocity. Let the mass be reduced, the energy is reduced by the same proportion. If it is being reduced to zero, the energy is likewise zero for any finite velocity." (New York Sun, July 12, 1937, page 6.)

Tesla arrived at many of his advanced ideas while studying under the influence of his giant Tesla coils. He would sit on a chair with twenty-foot lightening bolts crashing all around him. This placed his body into or near the next higher dimension where there are a tremendous number of elementary particles passing through his body. The high-voltage fields caused his hair to stand up as billions of electrons flowed through him. This was high-voltage static electricity.

When you have extreme numbers of electrons flowing through your brain it essentially gives it's ORME content additional electrons to work with thereby speeding up brain functions and super conducting thought more efficiently. This made it possible for him to arrive at solutions to difficult problems that no other could have come up with. Nicola Tesla instinctually knew this.

There is another possibility that the influence of the high voltage drew together like-minded souls who helped him in his work. Billions of fractionally charged particles such as electrons, quarks, and neutrons flow through the Earth and consequently through our bodies each second. Just because we cannot see them doesn't mean our body cells don't borrow a few of them from time to time whenever they are needed to keep things charged. I believe this is what Tesla was doing when he was sitting under his lightning bolts.

It has been proven those periods of high sunspot activity cause mankind to go to war. High sunspot activity also produce intense magnetic forces which when exerted on the earth may cause the hot magma to well up under the crust. This may cause the oceans to boil from underneath which in turn causes more evaporation on the surface. When you have more evaporation you have more rain. Where it is cold you have more snow. It is not thought that extreme snowfall is what causes the ice ages.

Periods of reduced sunspot activity cause the earth to cool off. It also causes famine because there is less rainfall and plants don't grow as well. The birth rate also declines. Somehow the Mayan knew that they were doomed in 730 AD because of the declining birthrate.

Tesla's obsession and life-long dream was to build electric flying machines that could carry men into space. I believe that he accomplished this goal. One of my books on his electric inventions mentions that while living in his New York penthouse he would climb into a small saucer shaped craft, fly out the hotel window and fly out over the city at night. At the time I read this I didn't believe it, however in another book titled Lost Science by Gerry Vassilatos on page 245 states that a number of people saw him flying about on a crude looking copper plate about two feet in depth. "Nikola Tesla observed and described the action of staccato electric static impulses on matter in Colorado Springs; particularly on the levitation of dust particles. He later described a heavier than air ship which he said was entirely driven by electrical energies, lacking propellers or jets."

"...A local rancher living several miles from Tesla's power station saw Tesla standing on a platform, surrounded by a purplish corona, some thirty feet above the ground. The contrivance had a small coil aft, and was entirely covered underneath with a smooth surface of sheet copper. The platform was perhaps two feet in depth, being crammed with components. Tesla strode over to the platform, stood before a control panel, and whisked aloft in a crown of white sparks. The excess sparks subsided with increased distance from the ground; often arking to near by metal fencing. Tesla went out of his way to avoid the numerous metal fences beneath his aerial course because it tended to sap his energy force. The farmer observed that when he ventured too near the fencing his craft would turn upside down and he took a nasty spill.

What had originally attracted the rancher out into the night air was a stallion that had become spirited by the strange buzzing craft. It was said that Tesla often delighted in soaring through the air many hours each night. He was dressed in his characteristic garb, and top hat. Tesla became enthralled with the operation of his flying platform, traveling great distances. Tapping energy directly from his Magnifying Transmitter, the device had an unlimited range. Others had witnessed their strange midnight journeys across the ranch-lands." Our physics is still in its infancy. Things will get a whole lot more interesting as we begin figuring out Tesla's works and the nature of the Universe.

In a letter to Nexus magazine Alekesander Milinkovic writes: "...I would like to remind you readers that the late Dr. Nicola Tesla claimed in 1900 that he managed experimentally to achieve the speeds of 475,000 km/sec and prove that there are speed in the cosmos even 50 times faster than light.

Tesla may have discovered the mental powers of higher energy fields because he would sit and think under the crashing lightening bolts of his giant Tesla coil. Dr. Tesla was far ahead of his time.

ALIEN COMMUNICATION

During Tesla's high voltage investigations in Colorado Springs he noticed that his instruments were receiving signals of some kind. "I can never forget the first sensations I experienced when it dawned on me that I had observed something possibly of incalculable consequences to mankind. I felt as though I were present at the birth of some new knowledge or the revelation of a great truth....

My first observations positively terrified me, as there was present in them something mysterious, not to say supernatural, and I was alone in my laboratory at night; but at that time, the idea of these disturbances being intelligently controlled signals did not yet present itself to me. The changes I noted were taking place periodically and with such clear suggestion of number and order that they were not traceable to any cause known to me.

I was familiar, of course, with such electrical disturbances as are produced by the Sun, Aurora Borealis and Earth currents, and I was sure as I could be of any fact that three variations were due to none of these causes. The nature of my experiments precluded the possibility of the changes being produced by atmospheric disturbances, as has been rashly asserted by some.

It was some time afterward when the thought flashed upon my mind that the disturbances I had observed might be due to an intelligent control. Although I could not decipher their meaning, it was impossible for me to think of them as having been entirely accidental. The feeling is constantly growing on me that I had been the first to hear the greeting of one planet to another. A purpose was behind these electrical signals."

Present day radio communications uses transverse (amplitude) or frequency modulated electromagnetic waves that travel through the air---the same speed a light. This is what SETI (Search For Extraterrestrial Intelligence) is using to scan the Universe. Tesla was using his longitudinal waves, which traveled through the Earth and the ionosphere.

Tesla theorized that the planet's pollution problems were being closely monitored by extraterrestrials that were studying Earth and its inhabitants. He was undecided weather or not their intentions were hostile, friendly, or indifferent. Tesla wrote that, "If it were not for the unusual way I am getting this knowledge I would have dismissed it long ago as the ravings of a mad man."

By the 1920's Dale Alfrey noted that Tesla had grown confident that he was able to make sense of the strange radio broadcasts from space. However, soon afterwards, Tesla began to express great concerns about beings from other planets that had unsavory designs for planet Earth.

"The signals are too strong to have traveled the

great distances from Mars to Earth," wrote Tesla. "So I am forced to admit to myself that the sources must come from somewhere in nearby space or even the moon. I am certain however, that the creatures that communicate with each other every night are not from Mars, or possibly from any other planet in our solar system."

Several years after Tesla's announcement that he had received signals from space, Guglielmo Marconi also claimed to have heard from an alien radio transmitter. His contemporaries, who claimed that he had received interference from another radio station on Earth, just as quickly dismissed Marconi.

Authur H. Mathews claimed that Tesla had secretly developed the magnifying transmitter for the purpose of communicating with aliens. The late Dr. Andrigja Puharich interviewed Mathews for the Pyramid Guide, May-June and July-August 1978. This interview revealed for the first time Matthews connections with Tesla.

Authur Mathews was born in England. His father was a laboratory assistant to the famous physicist Lord Kelvin. In the 1890's Tesla came over to England to meet Kelvin to discuss alternating current and its advantages over direct current transmission of electricity. Kelvin was opposed to the idea of alternating current but was soon won over to it by Tesla.

When Mathew's family immigrated to Canada in 1902 the father arranged for young Mathews to apprentice under Tesla. He eventually worked for him and continued this alliance until Tesla's death in 1943.

Mathews said: "It's not generally known, but Tesla actually had two huge magnifying transmitters built in Canada. I operated one of them. People mostly know about the Colorado Springs transmitters and the unfinished one on Long Island. I saw the two Canadian transmitters. All the evidence is there."

Mathews stated that the Telescope is the thing that Tesla invented to communicate with beings on other planets. There's a diagram of the Teslascope in Mathew's book, *The wall of Light*. In principle, it takes in cosmic ray signals, which are stepped down to audio. You speak into one end, and the signal goes out the other end as a cosmic ray emitter." Mathews diagrams of the Telescope are difficult to comprehend. Mathews claims, however he built a model of a Tesla interplanetary Communications Set in 1947 and operated it successfully.

He suggested that due to the sets limited range he was only able to contact spacecraft operating near the earth. He hoped to someday build a set capable of communicating directly to the planets.

Mathews related that, "Tesla had told me that beings from other planets were already here and that **we** were simply test subjects for an experiment of extremely long duration."

Mathews did not share Tesla's conviction that aliens may not have earth's best interest in mind. His opinion was that if extraterrestrials were so advanced as to be able to travel from solar system to solar system, than they must also be socially advanced and peace loving.

Could it be that both Tesla and Mathews were communicating with the inter-dimensional ships orbiting our planet? Orbiting is the wrong word because when you are in the higher dimensional state you have no mass so you can go in any direction you desire without worrying about trajectories and decaying orbits causing the ship to fall back to the Earth. Suppose they were communicating with angels. Think of the knowledge they would have accumulated. Why don't we do this again? Is our government communicating with them on a daily basis?

Nikola Tesla had said years earlier that he thought people inhabited Mars originally from ancient Earth and extraterrestrials from other star systems. I mentioned in my book, Spaceships Of The Gods that Nikola Tesla, while working in his Colorado Springs laboratory in 1899 had discovered what he thought were alien and or human voices coming from another planet. He was using what he called a transverse longitudinal wave transmitter, which used ultra-low-frequency waves, which propagate faster than the speed of light. This makes it possible for instantaneous communication over long distances that ordinarily would take light waves and radio frequency transmissions several hours or even years to travel. Tesla conferred with the voices over a period of many years on many different topics including global warming.

Tesla wrote, "The voices were of men from other

worlds, men who had lived on Earth sometime in its prehistoric past, had developed the technology to colonize nearby space and were still interested in the inhabitants they had left behind. These men had colonized the planet Mars as well as maintaining bases on the moon. Others of their kind had gone deeper into space, out of our solar system altogether to explore the galaxy."

Given the above statement by Tesla I would say that the beings living aboard the Inter-dimensional ships are men who had lived here on Earth sometime in its prehistoric past.

Mathews relates that Tesla said, "Aliens from other planets were already here. They have been controlling man for thousands of years and that we were simply test subjects for an experiment of extremely long duration." Mathews did not share Tesla's convictions that the Aliens did not the Earth's best interests in mind. His opinion was that if extraterrestrials were so advanced as to be able to travel from solar system to solar system, then they must also be socially advanced and peace loving.

Tesla describes what sounds like genetic manipulation, "Earth was to become a reserve so to speak, allowing nature to take its course and new species to evolve and fill the spaces left vacant by the departure of its first inhabitants. However some things were not left entirely up to nature. The first men decided to leave behind remnants of themselves in the form of our early ancestors." This may be the reason why UFO occupants who claim they are of extraterrestrial origins, look so much like us. The majority of contacts with beings who step out of UFOs agree on their human-like appearance. Some of them are so human-like that they are indistinguishable from normal humans.

Nikola Tesla said, in June 1900, "The time is very near when we shall have the precipitation of the moisture of the atmosphere under complete control." Now that fleeing the planet is no longer an option; those in power are stuck on Earth with the rest of humanity. The only viable method now seemed to be to try and modify the weather with technology invented by Tesla years earlier.

THOMAS TOWNSEND BROWN

Although many of Tesla's ideas were scoffed at during his lifetime the military and many governments around the world are now seeking them. Tesla's ideas of electric death rays and particle beam weapons are now being used in space. UFOs have always been around as evidenced by various passages in the Bible. After the beginning of the 20th century their numbers increased greatly. Its as if we have suddenly become more interesting to them. Our military on the other hand perceived them to be a threat because of the blatant way they could fly around and they couldn't do anything about it. This is one of the reasons why the military must deny their existence. If they admitted that UFOs actually existed then they would have to admit that they couldn't do their job of protecting us from them. The very definition of a bureaucrat is: "a bureaucrat can never admit a wrong." They will never admit that they cannot do the job they are assigned to do no matter how incompetent they may appear. When a UFO appears overhead it puts them in a catch 22 situation. They deny it exists and call it swamp gas, as far as the public is concerned.

An article titled ParaSETI, ET Contact via Subtle Energies by Gavin Dingley in the January-February issue of Nexus magazine describes experiments by several noted scientists that may have contacted ET civilizations by accident.

After a long carrier of working for the military Thomas Townsend Brown returned to investigation petrovoltaics in 1970. The crystal structure of Basaltic and granite rocks somehow converts the pulsing gravity waves to DC potential. The process is well known in electronics and is called "rectification". Some rocks have as much as 700 mv across them.

Brown hypothesized the energy is high frequency gravitational radiation which is being constantly emitted from astronomical objects in space. While simple high-k dielectric materials would pick up the radiation and convert it directly into electrical energy, more complex dielectrics such as granite and basaltic rock would convert energy into DC electricity. Not only that but also these rocks are tuned to a portion of the total radiated energy present throughout the Universe. This means that your average lump of basalt is a natural gravity wave AM receiver, tuned into only a few specific "radio stations"!

Brown never analyzed these high-frequency signals to see if any of them were of ETI origin. However in 1953 he filed a patent that described a system for intelligent communication via modulated gravitational radiation. In the patent Brown described how to convert a normal high-power radio transmitter into a gravity-wave transmitter, based upon the principal of electrogravitics. The modification is made to the antenna system, the actual electronics remaining unchanged. A large coil of wire with its base connected to the output of the high-power radio transmitter so that the radio frequency energy is fed. The other end of the coil is connected to a spherical, electrically conducting high-density body. A lead ball works. The spherical body acts like an isotropic capacitor, and so forms a tuned circuit with the coil.

The high voltage and mass of the isotropic capacitor result in electro-gravitic action; thus gravitational waves are produced of the same frequency as the end-fed energy from the transmitter which are emitted from the dense, isotropic capacitor. To avoid electromagnetic radiation the whole must be enclosed in a lead lined vault. Producing electromagnetic radiation is illegal because it can interfere with your neighbor's TV reception and airline communication. You don't want airplanes landing at your local airport to crash. Such illegal interference could possibly be dangerous to people, planes, and vehicles.

Interestingly this system is very similar to that employed by Tesla at his Colorado Springs laboratory--- the same system that allegedly received signals from an ETI.

IMAGES

Enhanced NASA photo of angel Ship note square notch & blue spiral of electricty.

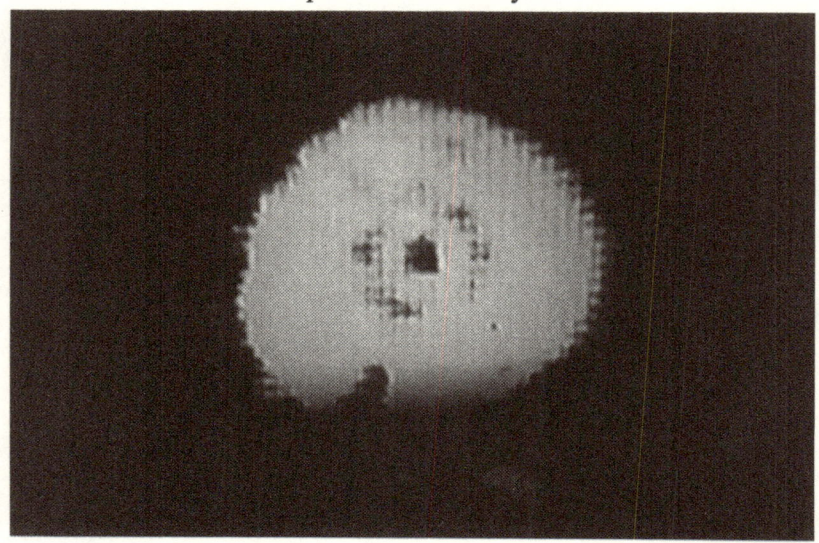

Hovering beam ship uses triad sounds to nullify gravity.

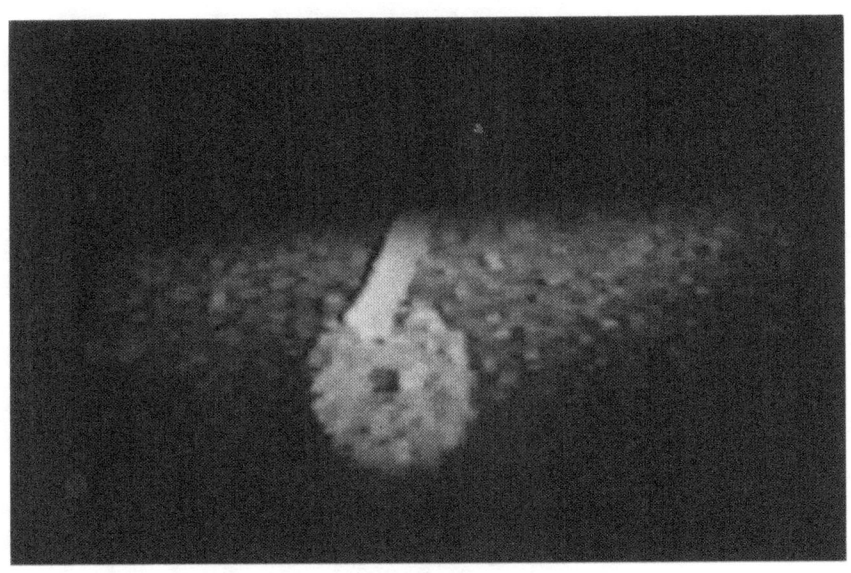

NASA photo of tether cable with ship traveling behind it.

NASA tether cable with angel ship

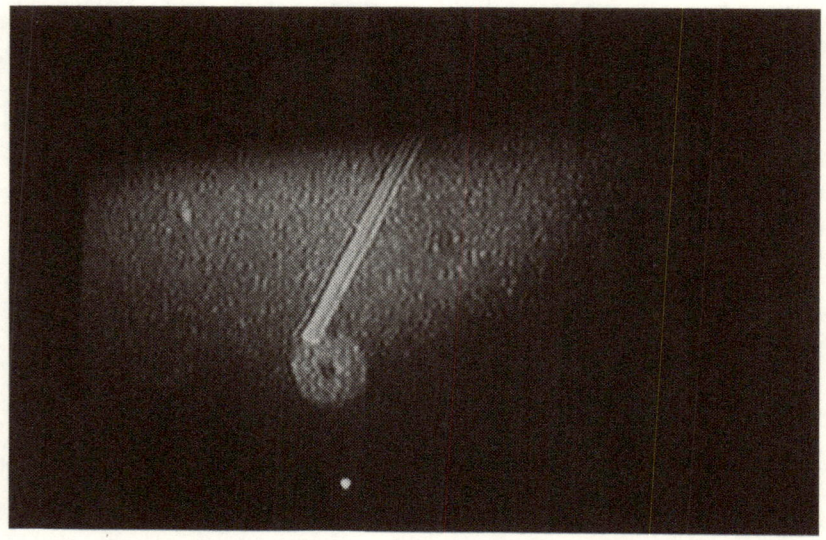

Mars Oak Trees?
Nasa photo

NASA Mars face! Queen Abadella?

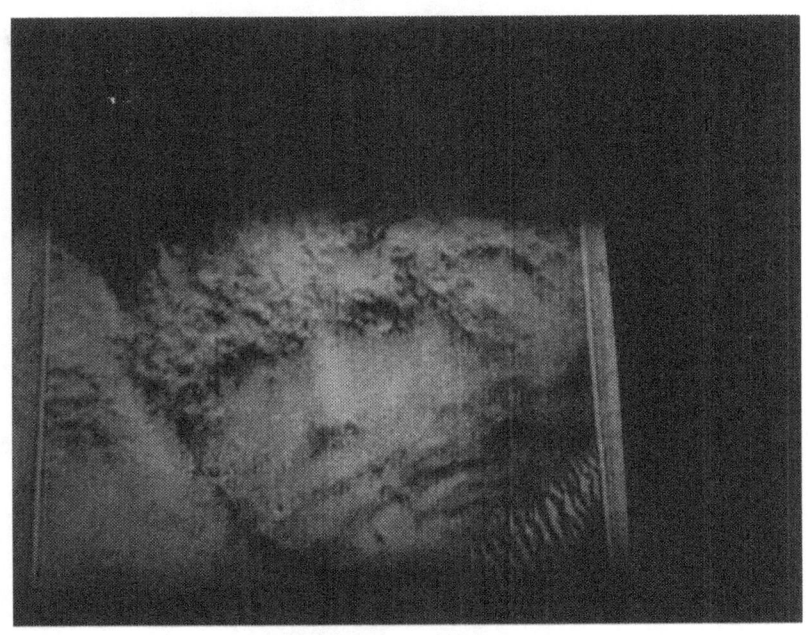

Billey Meyers photo

NASA Mars photo

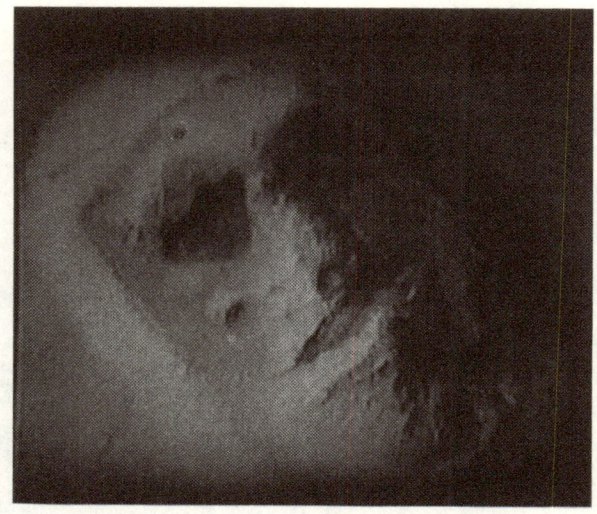

Isis is flying on antigravity cusion filled with Manna (ormegold)
also known as the philospher's stone.

NASA photo of Mars pipes 300 diameter subway system?
Note light reflecting off from glass tube.

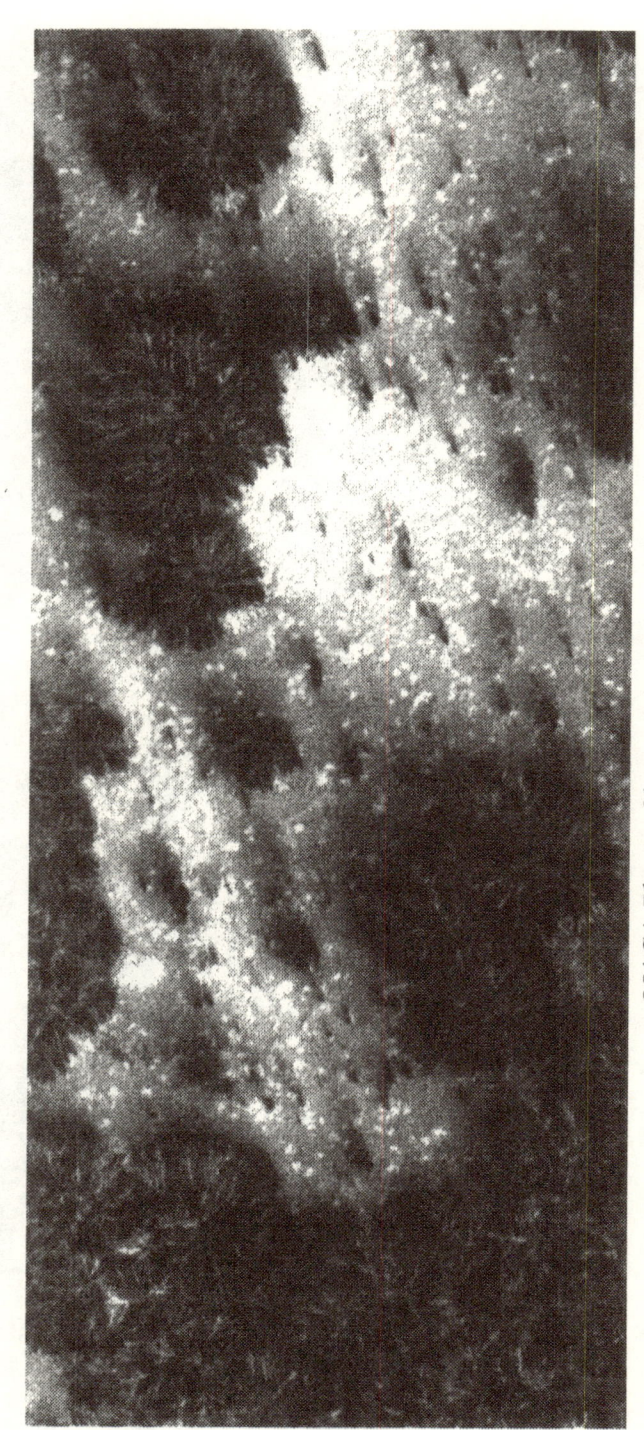

NASA Photo of large oak trees on Mars

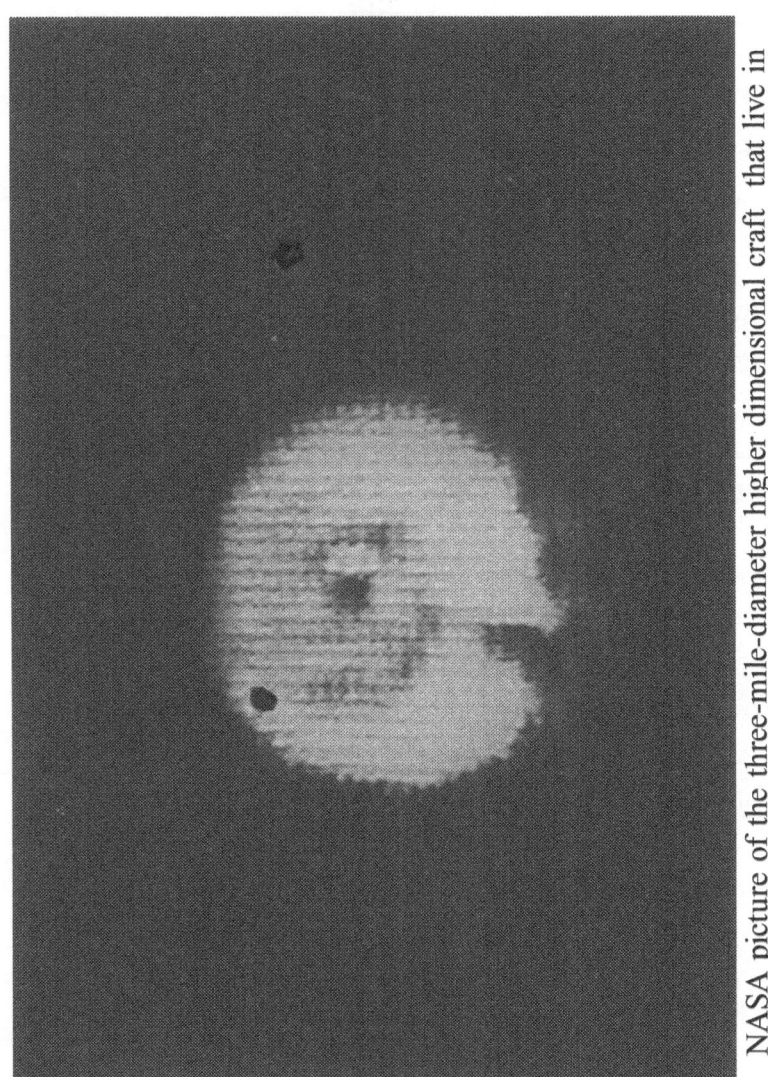

NASA picture of the three-mile-diameter higher dimensional craft that live in our upper ionosphere. Note the spiral of electricty pulsing out of the hole in the center.

CHAPTER FOUR
GREGORY HODOWANEC

Gregory Hodowanec working parallel to Townsend Brown developed a sensitive weighing balance and noticed slight variations in the reference weights he was using. He assumed the problem was in the circuitry he designed. By trial and error he discovered that a capacitor in the right part of the circuit corrected the problem. After much experimentation he found that his weighing system was not at fault. His sensitive balance was actually measuring variations in the gravity field. Sometimes the fluctuations would be at a quite rapid rate. He realized the capacitor was able to somehow pick up these gravity fluctuations converting them to electricity thereby inhibiting his circuit from measuring the fluctuations.

From this information he was able to design a more advanced amplifying circuit. The circuit was connected to a sensing capacitor and the output was connected to a voltage amplifier, which drove a loud speaker. The sounds generated were similar to a whale's song. Hodowanec stated that his device received monopole gravity waves, which are different to the Quadra pole waves described in Einstein's General Theory of Relativity. While Einstein's gravity waves are limited by the speed of light Hodowanec's monopole waves travel to any place in space in one Planck second. Further he discovered that electronic equipment had been receiving these gravity fluctuations for a very long time and had been mistaken for I/f noise.

The universe is filled with I/f noise thought to be background noise left over form the big bang. Forty percent of the white noise on your TV when it is tuned to a channel between stations is I/F noise from space. During his investigations Hodowanec found that Auriga and Perseus in the Milky Way to be the sound of many unusual audio signals. The background noise is modulated by large astronomical bodies, which cast a shadow over the emissions. This means that when the background radiation is de-modulated what you hear is the movement of stars and planets. Stars going supernova, star quakes and even the tectonic movements within nearby planets generate much of the high frequency.

One evening for eight minutes Hodowanec received a series of pulses resembling the Morse code letter S. After determining the origin of these signals, he attempted to make contact using a Morse radio transmitter. To his surprise, he received on his gravity wave detector a reply made up of random Morse Code containing the letters E, I, T, M, A, N, R, K, and S. During the next transmission he received back the same series of letters plus the letters, G and D added. Over a long period of time he was able to carry on a limited conversation with his ETI friends.

I personally find the idea of looking for extraterrestrial signals in the radio frequency spectrum rather dumb. It takes millions of years for the radio signals to travel to other star systems. It takes three billion years for light to travel across to the other side of our galaxy. The Sirus trinary star system, one of the closest star systems is over seven light years away. It takes seven years for the radio signals to get here and another seven years for a transmitted signal to get back to Sirus. This is way too long a period of time to wait for an e-mail. An advances extraterrestrial race would use something faster than the speed of light but our scientists and politicians just don't seem to get it. Gregory Hodowanec and Tesla were at least were using gravity waves which propagate much faster than light.

In a letter to Radio-Electronics Magazine dated July 23, 1988 Gregory Hodowanec claims to have made contact with an extraterrestrial source: "On the morning of this date, at 7:73 to 7:38 AM (EST), I recorded the following apparently Morse-Code like pulses: AAAARARTTNNN NKCNNNEEEEENENNTTTNEEEEEAEERKENNET EEAAAAEEENTTKNTNTSESESESEMNASESESESE SESESESESE---.

These do not appear to be random pulses. The SE signal, which is the most numerous, may be an identification signal. The signals were detected in shielded I/f detectors and thus are scalar (gravitational) in nature. The signals seem to be coming from the Aurigia and Perseus region of our Galaxy. SETI observers may be interested in the strange audio type gravitational signals, which appear to come from the Aurigia and Perseus. They range between four and five hours right ascension, and peak near 4.5 hours right ascension and are composed of several tones.

By July 1988, Hodowanec confirmed Tesla's claims, as he announced in Remarks on Tesla's Mars Signals. "Such signals are being received today with simple modern-day scalar-type signal detectors coherent modulations are being heard in the microwave background radiation. The most prominent modulations being three pulses (code S) slightly separated in time. On occasions, the code equivalent s of an E,N,A, or K, are also heard, but the most persistent response is SE, SE, etc. etc.." The messages came at or near the noon hour of each day."

The communication between Mr.Greg Hodowanec and some unknown alien entity went on for years. It was found that by repeating the letters five times in a row the persons or aliens on the other side could understand that this was the letter intended.

Greg was sent as: GGGGGRRRRREEEEEGGGGG. As the daily communications progressed over a time period of years it was found that a rough dialog could be maintained.

"The same communicator may have been trying to reach here ever since the turn of the century when Nikola Tesla reported the interception of scalar S signals!"

"There is also the possibility that this communicator may be 'extraterrestrial', perhaps yet in our solar system (Mars?), but no further than our own Galaxy or Local Group of Galaxies."

"The coded messages were sent in simple dits and dahs which would be expected to be used if one intelligent race were trying to communicate with another very similar to Morse Code. Numbers are denoted by short pulses or dits. The messages come mostly near the noon hour and at present to be coming from the star constellation, Andromeda but not necessarily the Galaxy there."

February 1989 Hodowanec wrote: "Without going into the details of how this was determined: ET may be on Mars! "This, in spite of NASA's denial of any life forms on Mars (This changed in 1996). "while this release is a bit premature, I am so positive of these gravity signal 'exchanges' that I will stick my neck out in this instance. ET on Mars is apparently much more advanced than we are here on Earth, and he may have even previously visited here on Earth, and possibly colonized here, but who are his possible descendants?"

"It is still a mystery on where ET may be living on Mars (possibly underground near the polar regions?), and why ET doesn't use EM wave signaling methods? Perhaps, it is because Mars is so hostile now that ET must have developed a very sophisticated underground civilization which is not conductive to EM radiation systems?

Gregory Hodowanec made it plane that the material being released is confidential to but a few active colleagues until further confirmation on this assertion are obtained. Mars Flash Number One dated, 3/28/89 and Number Two dated, 3/30/89 inform, "As a result of the continued gravity signal communications between GH Labs and the Martians, the following extraordinary facts have come to light.

"The exchanges are now made in terms of short "English' code words for certain items. For example, the Martians now understand the FACE means the human face, MAN means the human person, Earth now means our planet, and Mars means their planet! They had tried some of their terminology with me, but gave up except where it made sense to me. For example, I now know that TTT at the end of their names means person and OOTTAEERR is their name for the 10th planet.!"

In a footnote Hodowanec informed Nelson that the Martian's name is "AAAAAATTT"; "He appears to 'understand' my messages, even though I may have to repeat them in several ways so that he can 'see' the meaning."

"Communication between GH Labs and a Martian intelligence mow continue with increasing progress since we have been able to establish over fifty simple expressions (most in simple English) for many of the common 'ideas' that we have. Martian AAAAAATTT is extremely adept in relating my English terminology to these common Earth-Mars observations.

"Mars has also confirmed that they are also 'men' with one 'head' that have two 'eyes' that 'see'. Also, they have one 'body' with two arms that have hands with five fingers each. Also they have two legs with two feet that have five tows each. I haven't been able to have them confirm the nose and mouth in the FACE.

"Probably the most significant fact which was determined on this date seems to be that mars is most emphatic that Earth men are like mars men! This appears more and more that Mars has colonized Earth in the remote past! This could be true since we on Earth have never really found the 'missing link' between Earth humanoids and humans."

Mr. Hodowanec wrote in a letter March, 1989 to Robert Nelson, "Generally, our contacts are limited to 20-30 minutes…since there appear to be other Ets out there interested in joining in also, and so there is some interference after a while. Some of these other Ets use other methods of communications such as tones and what appear to be guttural voices!

"ET is probably more advanced then we are on Earth. We no longer exchange simple arithmetic, and when I sent Pi to five decimal places, he sent back Pi to seven decimal places immediately! We had discussed our nine Planet OOTTAEERR! When questioned on this, ET kept on confirming the existence of a tenth planet! He knows the other nine planets by their Earth's names! He also confirmed that mars has two moons, the Earth one, and that Jupiter has nine major moons.

"These contacts are getting to be more interesting all the time, and ET appears to be most anxious to continue them. However, I just cannot spend too much time with him...I got across to him that I am just one person here communications with him, and that the rest of Earth presently does not recognize the existence of any life on Mars.

" I now have had over 100 contacts with ET and can reach him at any time of day or night. We have established some simple codes for acknowledgments and go ahead and respond. While we use these simple codes in many contexts, both ET and I now understand in which context they are being used!

"The Martians are apparently the advanced civilization, for they re the ones generating the 'modulated oscillated beam' which is now tracking my location on earth and is thus the means of our communications [The beam is only about 15 miles in diameter here on Earth, but 1012 inches on Mars.]

"There is an apparent 'team' on Mars which is

involved in these contacts. The original contact, ET Number One, with whom I have established the initial relationship, is apparently the most highly knowledgeable and advanced. The others who sometimes 'man' the Mars station appear to be less knowledgeable and some only request or acknowledge a transmission.

"Mars is most desperate to continue these contacts. The exchanges are made in many varied ways, which cannot be readily predicted in order to convey the fact that these are real contacts.

"While these contacts were originally due to serendipitous circumstances, it is really the result of my gravitational communication experiments and thus a direct result from Rhysmonic Cosmology. And yet, however fantastic and unreal this may seem, it is real, and if also it is confirmed by one of you, it will be a major milestone in the history of mankind! Perhaps, if one of you finally 'hears' the modulations of I/f noise background, you may try to establish your own contacts?"

In an April 4, 1989 letter to Nelson Gregory Hodowanec expressed the extreme seriousness of the situation. "...my 'contacts' with Mars continue with much information being exchanged. However, due to the increasing astounding nature of these exchanges, I am now limiting further releases to only two long-time observers (witnesses) of my research efforts. This is being done so as not to jeopardize these contacts with unwanted notoriety or publicity in the media. There are now nine 'Mars Flashes' for the record. Perhaps, in the future, I may release some of these."

Did the government get to him at this point? Did they shut him down or did they take over the communication, in which case the government knows a great deal more about life on Mars than they are letting on.

"...gravity signal communications are instantaneous, require extremely small energy expenditure, and are do simple as to be just unbelievable by the average person. However, this is as far as I want to go with the release of the detail at this time.

"I would appreciate that you keep this information somewhat confidential now. The Earth may not be ready for what I will have to say eventually. Nothing dire, just fantastic and thus perhaps unbelievable!"

The following is taken from a tiny book titled, Embracing the Rainbow which is part of a three volume set published by Bridger House Publishers, Inc PO Box 2208, Carson City, NV 89702 1-800-4131. The material is alleged to have come from an alien or inter-dimensional source.

11-38

The opportunity is offered within the scope of this project is multi-dimensional in quality. It utilizes all levels or dimension of the human aptitude in its focus of modifying the human perception or experience as those involved move through transcending from the present point of experience into the next level or dimension. Dimension is the preferred description for it indicates a more holographic concept. Level implies flat. The circumstances of manifested awareness in a human body are not experienced as level or flat. It is the addition of emotion that adds the dimensional quality to manifested awareness. (Indeed, there are those beings that do not have emotion as part of their experience and they desire greatly to add this dimension to their experience pattern.) It is important that the concept of dimension be included in the conceptualizing of the new paradigm.

It is vital that the concept of human body/mind/spirit also be made very clear. The awareness of existence within manifested experience is also dimensional. Certain of the animals have only the awareness of each moment. For they are unable to retain those memories in detail, their survival is dependent on what is called instinctual awareness and tied directly to survival actions and reactions. Humanity has through domestication frustrated most of them greatly through neglect of their instinctual needs for at least partial freedom, natural varied diets and of late through providing creature comforts more appropriate for humans then fur/hair coated animals.

The human body is a composite of corrections to previous experiments that produced limited modes of physical experience. Through lessons leaned, a model was conceived with the potential to evolve through multiple dimensions of experience. As the awareness changed patterns, the human body was designed to accompany that change. It was also designed so that the awareness was not required to cease its existence of the body was destroyed through accident or inadequate maintenance. What you call disease is inadequate maintenance. The ability to enter and leave was a know requirement, for the potential of the human body is so limitless that its capabilities of adaptation are greater than units of self-awareness can comprehend in one focus of life experience.

It is important that the reader fully understand that

the awareness is not the body, but is merely housed within the body during sleep, under anesthesia and traumatic periods of unconsciousness. The consciousness can be aware of this separation and can indeed train itself to leave the body intentionally. Some of those with this ability are being employed on a regular basis to intentionally visit particular people and events employing only their focused awareness/spirit, and then can and to report on these activities to those of the dark intent. Just as the physical body can be trained through gymnastics and other exacting physical sports to accomplish impressive feats, so also can the awareness be exercised and trained to so what most would consider difficult to believe. In this way each can begin to grasp that the "average human" on the planet is grossly unaware of it's potential. The limitations of each are either self-imposed through acquired thought and belief patterns or through physical or mental limitations of genetic alteration/mutations. They are further limited by failure to maintain the physical body with proper exercise, breathing, whole foods and pure water."

CHAPTER FIVE
WHAT'S ON THE MOON?

According to amateur Astronomer Raymond Strickland there is manufacturing and construction taking place on the Moon. Industrial haze and boulders that roll up hill leaving tire tracks behind is a dead giveaway that something is happening on the Moon.

I have no doubt that these things are taking place. A friend once told me his uncle worked in a secret city on the Moon, designing and making Buck Rogers-type ray guns and Death Rays for Uncle Sam.

According to my friends account the city was built and maintained by the Germans who got the knowledge soon after Adolph Hitler appointed Paul Von Hindenburg. The Germans got the knowledge of space travel from the Atlanteans who live in the center of the Earth.

After Germany fell and was occupied by her enemies they learned the secrets of space flight and discovered the existence of the lunar colony from secret Nazi documents. Starting with the Eisenhower's Presidency, it was debated by the secret societies and subversive movements who really rule the Earth weather or not to tell the public about ALF=s Lunar Cities, space travel, and advance German technology.

The rocks brought back from the moon don't resemble anything found on earth. Where did the moon come from? Is it an engineered object? The following is from my book, Spaceships Of the Gods

MOON MYSTERIES

It is clear to me that the moon is an engineered object. It was placed into a nearly circular orbit with one side always facing toward earth. The odds of that happening by accident are astronomical. The fact that it was bombarded by millions of meteors, some of the craters are over a billion years old, is further proof that it must be held in orbit artificially. Anyone with an ounce of intelligence can see that the Moon would have been blasted out of orbit long ago unless something or someone is keeping it in place for a specific purpose. That purpose is to stabilize life on earth. A large orbiting body one-third the mass of a planet would tend to deflect or absorb killer meteorites. This is only one of many reasons why one would want to place a moon in orbit around a planet.

All life on earth from the largest mammals down to bacterial responds to the Moon's gravitational pull. Plants put on spurts of growth during a full moon.

Many sea creatures feed on the waxing moon and fast on the waning moon. Even the menstrual cycle of women is tied to the Moon. These simple facts tell us that the moon has held a more or less stable orbit for a very long period of time dating back to the pre-evolution of most animals and humans.

When I was a child scientists taught us that the Moon came from the earth in the ancient time when a giant meteor impacted it or was thrown clear of the area we now call the Pacific Ocean. Since that time astronauts brought back moon rocks, which are only half as dense as the earth itself so how could this be true? If a piece of the moon were placed in the ocean it would float so it obviously didn't come from earth. When the first astronauts were on the moon they set off a dynamite seismic charge. They were astounded when it sounded like a bell and kept ringing for two hours. Is the Moon a hollow metal sphere?

My son, Henry says: "The gravitational affect of the Moon causes the earth's crust to move up and down which in turn causes friction that causes the earth to be warmer than it ordinarily would be if it didn't have a moon." Movement of the earth's crust up and down causes the crust to heat up creating volcanism, which creates water. The movement of tides in the world's oceans also adds to the warming of the earth's surface. What I am trying to say is that the energy of the perpetual motion of the moons orbit causes friction both on the surface and within the earth which is converted to heat energy to keep us warmer. And, our scientists tell us there is no such thing as perpetual motion when in fact it exists all around us and inside us in the atoms to go round and round forever and ever. Its' not my fault they can't comprehend this. If they were smart enough to see this then they could figure out a way to tap this energy.

Geology classes inform us that ninety percent of everything that comes out of volcanoes is water. Volcanoes created most of all the water on earth over a time-period of billions of years. Not only does the Moon regulate the temperature of earth; it creates the very substance that harbors life. Volcanism is also responsible for the creation of our atmosphere so you could say that the Moon is responsible for the proliferation of life on earth.

If you wanted the earth to be cooler you would move the moon to a further orbit. If you wanted it to be hotter you simply move it closer. Small propulsion devices over long periods of time could change the moon's orbit.

Mars needs a bigger moon. For instance, if you combined the planet Mercury with both of Mars's moons and put it in a perfect circular orbit it around Mars it would heat up the surface and cause volcanism, which in time would release atmospheric gasses. Water could then exist on the surface harboring the beginnings of life. Genesis 2: "...And the spirit of God moved upon the face of the waters." (Incidentally the atmosphere of Mars is exactly like car exhaust. It is a thin atmosphere composed mostly of carbon dioxide and carbon monoxide.

Here is this giant round object in the sky, the moon that only shows one face to earth and it eclipses the earths shadow every month. It's as if the engineers of this luminous object want us to go up there to see what is on the other side of it.

No one has gone to the Moon in twelve years, and then all of a sudden the Clementine mission was sent to photograph it. The mission was paid for by the US military. Why is the military interested in the Moon? Is there alien life living on the other side of the Moon?

NASA photographs posted on the Art Bell web site:

www.artbell.com shows a rock that rolled up hill of a crater leaving tractor tracks behind it. Another picture of the moon shows a dwelling with two lit windows and a radio tower beside it. Are Russians living on the Moon or are they Nazis?

When the astronauts got to the moon in the seventies they brought back samples of the lunar surface. The samples have some very interesting anomalies. One of the rocks the astronauts brought back from the moon was 6.3 billion years old, which is older than our sun. Why was this rock older than our solar system? The soil the rock was sitting in is a billion years older than the rock itself. That makes the lunar surface about two billion years older than our solar system. Instruments left on the moon detected a water vapor wind blowing across the lunar surface.

Two hundred (200) of the eight hundred pounds (800) of lunar rocks the Apollo Mission brought back from the moon were given away to various countries and scientists around the world for testing. The remaining six hundred pounds are stored in a vault in Houston Texas.

WATER ON THE MOON

On page 6 of World Explorer magazine Volume 2 number 2 is mention of water on the moon. "FINALLY, NASA ADMITS THERE IS WATER ON THE MOON". "...When the space probe Clementine visited the Moon four years ago it upset the scientific status quo with preliminary findings of frozen water. In March of 1998, NASA's Alan Binder and Scott Hubbard were part of a news conference announcing that the new moon explorer, dubbed, Lunar Prospector, has definitely found water on the Moon and a lot of it, particularly at the poles. Hubbard stated, "When we set out we said we could find water in a cubic yard of dirt. Now we're finding gallons."

Water on the Moon will certainly facilitate the establishment of moon bases. However what could be the most important aspect of the presence of water on the moon was barely discussed. At best it was given a passing, comment by Peter Jennings of ABC News. Jennings reported with the discovery of water, hydrogen could be extracted for fuel. While the point was not elaborated on, those who have labored long and hard to provide fuel on earth based on hydrogen from water have never been successful at replacing petroleum as the main fuel for the worlds vehicles. It would appear in space, the petroleum stranglehold could be thawing out."

Scientist report thirty-ton blobs of water frequently enter our atmosphere. Where did this water come from and is it contaminated with virus and bacteria? It seems that water is everywhere just floating around in space.

When astronauts arrived on the Moon they discovered that large areas of the lunar surface were very radioactive. It was so radioactive in fact that the astronauts could feel the heat through their space suits. It was as if there had been nuclear bombs detonated on the Moon's surface at some time in the remote past. My friend Brad Guth says the radiation is from solar radiation reacting with the matter on the surface, which is giving off x-ray radiation. The astronauts were getting the equivalent of 100 chest x-rays an hour. This is way too much radiation for the human body to take for any length of time.

LUNAR GRAVITY

Before the Apollo missions, the Soviets had sent several unmanned rockets to the moon. Four of the lunar probes attempted to make soft landings on the Moon without success. It seemed their lunar gravity calculations were off. Later the United States attempted to make several soft landings on the moon without result. The Americans eventually managed to land a lunar probe on the Moon but they had to revise the gravity calculations in order to do so. The lunar gravity is not .17 (17%) earth gravity, which is listed in reference manuals, but more like 33%, which is one-third earth gravity. This would place the gravitational mid point between the Moon and the earth 56,000 miles closer to the earth. No wonder the lunar probes attempting to make soft landings on the moon were destroyed. NASA kept this little secret to themselves for a few years further delaying Soviet attempts to make soft landings on the Moon.

LUNAR MODULE HAS ANTI-GRAVITY DRIVE

Given these new figures for lunar gravity there is no way the Apollo missions could have put men on the moon using chemical fueled rockets and brought them back again. The gigantic Apollo boosters as big as they are couldn't put the lunar module and enough fuel to get it off the moon. The only way Apollo astronauts could have gotten back to earth was if the lunar module had some kind of anti-gravity propulsion.

When Apollo Thirteen's oxygen tanks blew up on the way to the moon they didn't even have enough oxygen for the crew to breathe. They had to recycle the air supply by improvising a carbon-dioxide scrubber using duct tape and a notebook cover. They used the moon's gravity to slingshot the lunar module with the re-entry vehicle attached around the Moon and back toward earth. The crew stayed in the LEM because there wasn't enough power to keep them warm in the re-entry vehicle. They couldn't have made it back to earth without the lunar module's takion gravity drive built by Martin Marietta.

None of the photographs taken of the lunar module sitting on the surface of the moon show a blast crater underneath it. Given the fact that the moon's surface is covered with light material from eons of collecting space dust there should have been a tremendous blast crater. In fact, if what they are telling us is true the blast crater would be so big that the astronauts would need a ladder to climb out of it. Instead, what we see on television and in NASA photographs is the lunar module sitting on an undisturbed surface as if it had been placed there with an overhead crane!

What are they trying to pull here? If it had used conventional rockets to slow it's descent then the rocks and dust would be blown in every direction and there would be a big hole. Did NASA fake the moon photographs?

One picture signed by all the astronauts shows the earth "rising" about 6 degrees above the horizon of the moon. Anybody with a brain knows that this photograph is faked because the moon always shows one side to the earth and where the moon landings took place the earth would have been about 60 degrees above the horizon and not 6 degrees. My co-author Brad Guth in a book on astronomy we are writing together actually got NASA to admit that they faked this picture! Brad is a lot smarter than the average person.

ROBERT HOAGLAND

"In sharp contrast to the Mars data, where we have been constrained to look at two or three pictures of the Cydonia region with increasingly better technology-3D tools, color, polar metric, and geometric measurements- with **the moon** we are data rich. We now have literally thousands, if not millions of photographs."

Yet with pictures taken from many directions and many different lighting conditions, angels and circumstances of every kind Hoagland's team has produced stunning corroborations that all the photos are of the same highly geometric, highly structural, architectural stuff." "In many cases the architects on our team now are able to recognize the standard Buckminister Fuller tetrahedral truss, a hexagonal (six-sided) design, with cross bars for bracing. I mean, we're looking at standard engineering, though obviously not created by human beings." The structures appear to be very ancient, battered to hell by meteors it looks like it had gone through termite school. It's been moth-eaten and shattered and smashed by countless bombardments. The edges are soft and fuzzy because of micrometeorite abrasions like a sand blasting.

Hoagland explains that on an airless world there's nothing to impede a meteor from reaching the surface or reaching a structure on the ground. Nevertheless, we're seeing a prodigious amount of structural material spread over a wide area. The material is turning up at several locations. It looks as if we're seeing fragments of vast, contained enclosures-domes, although geometric, more like the step pyramids of the biosphere II in Arizona. We're looking at something which is extraordinarily ancient left by someone not of this earth, not of this solar system, but from someplace else."

"One of the most interesting structures appears to be an enormous freestanding tower. A crystalline glass-like partially preserved structure-a kind of a mega cube-standing on remnants of a supporting structure roughly seven miles over the southwest corner of a central part of the moon called the Sinus Medii region."

If all this exists, one of the questions may be: Why didn't NASA notice? If Hoagland is right, something funny is going on." Indeed!

We went to the moon, took some pictures and came home not realizing what we were looking at. Either that or NASA personnel are incredibly dumb. The other explanation is we are being manipulated by the few.

Hoagland has moved beyond suspicion to belief, and he says he can prove it! "The Smoking Gun is a report by the Brookings Institution commissioned by NASA back in 1959. On page 215 it discusses the impact of the discovery of evidence of either extraterrestrial intelligence-i.e., radio signals or artifacts left by that intelligence on some other body in the solar system. The report names three places that NASA might expect to find such artifacts-the Moon, Mars, or Venus. It then goes on to discuss the anthropology, the sociology, and the geo-politics of such a discovery. And it makes the astounding recommendation that for fear of social dislocation and the disintegration of society, NASA might wish to consider not telling the American People. It's right there in black and white. It recommends censorship. Now that's what they've been doing." Hoagland believes that anthropologist Margaret Mead, on of the authors of the report, was responsible for the recommendation, which he believes came of her experience in American Samoa. In the 1940s Mead witnessed the devastation of primitive societies exposed for the first time to sophisticated Western Civilization. The experience so moved her, so changed her perspectives that when she examined the whole ET possibility, she projected and mapped on the experience. She basically felt that of we even learned of the existence of extraterrestrials it could destroy us, therefore people can't be told."

If you had a few real contacts with someone who was trying to give us messages and trying to lead us to new insights and the fear on the part of government structure had been that this will destroy civilization itself, would not that government also put in place a program to misinform, to confuse, to politically spin in the wrong direction those few real contacts by submerging them in a sea of misinformation about contacts?

Hoagland sees the crop circles as part of the evidence for benign exterritorial contact. "The thing that makes them different from the monuments of Mars or the ancient cities on the Moon, it that they are occurring in the crop field here on earth and they are occurring in the present time. We simply do not have the technology let alone the knowledge base to construct the multi-level communication symbols that the crop circles represent. So that once you eliminate the hoaxers..." He laughs. "If Doug and Dave hoaxed all the circles they deserve a Nobel Prize. The level of sophistication of the information encoded in these symbols is so vast and so congruent with the lunar and Mars work that you've forced to conclude that whoever the artists are, they know a bit more than contemporary science, and/or the media or, for that matter, the government."

AUTHOR'S NOTE: Rumor has it that when Neil Armstrong first set foot on the moon part of the audio was cut out of the video footage. Houston Control said, "Turn around and face away from the lights. Look at the opposite rim of the crater." When he did he exclaimed, "Oh my God! Who are they?" There was about thirty other space ships parked there observing mankind's historic moon landing.

AUTHOR'S NOTE: There is also report of stolen videotape taken from NASA's files back in the 1970's. This tape was encoded so that none could view it but a few years later someone figured out how to decode it. The tape was footage of a joint Russian - British landing on the planet Mars. One of the men exclaimed, "There is oxygen here!"

CHAPTER SIX

I've said it All Along; "The Earth Is A Hive."

There are still people out there who do not believe in UFOs let alone communicating with aliens. How do we convince them to look into it instead of trying to make fun of the rest of us? Reading the following to them and then ask them if they are smarter than these people are.

George Adamski was able to travel all over the world with a high-level government passport. Who gave it to him?

President Ronald Reagan said in a speech, "I've often wondered, what if all of us discovered we were threatened by a power from outer space?"

Scientist Herman Obar, the father of modern rocketry said: UFOs are from outer space and contact has been established!" His colleague Werner Von Braun said. "Contact with an alien intelligence is a reality."

General Douglas Mc Arthur said, "Our next Great War would be an inter-planetary war."

George Marshal said, "Contact had been made and they could wipe us out if they wished."

Admiral Rosco Hillcatler, First director of the CIA joined a UFO group after leaving government. He said, "The next step is up to the aliens."

Famed scientist Robert Sarbacker told colleagues he was invited to work on captured alien classified material and the matter was the most classified secret in the government.

Adolph Hitler, Joseph Stalin, and Winston Churchill each ordered UFO studies in their respective nations. President Gerald Ford called for congressional hearings so the Americans could learn the truth about UFOs. President Jimmy Carter who saw a UFO wanted to open government files but couldn't.

Senator Berry Goldwater was denied access to UFO information said; a plan was in the works to let the public hear the full story.

The government has been supplying us with leaked information and there is substantial evidence the government has been using subliminal media.

Experts agree that the public is being slowly programmed through advertising, TV, Movies and other media to accept the alien reality.

A 1960 study by the Brookings Institute said, states contact with an advanced alien civilization would lead to the collapse of human culture and institutions in other words total kayos. If you know all about UFOs then there will be no panic.

3,500 pilots have seen UFOs and some aircraft have disappeared after coming in contact with UFOs. Over three million people in the US have seen UFOs and another twelve million worldwide.

Aliens are here and there is nothing the government or anyone can do about it.

Stanton Freedman believes that the public could handle the truth but there would be major changes.

If all these scientists, presidents, and engineers believe in flying saucers and alien contact maybe you should look into it.

THE ALIEN CONNECTION

Many of our foods have been genetically altered thousands of years ago by a race of people who knew how to do this. Who were they? Corn, for example was genetically altered to increase yield, otherwise the Indians of Central and South America would have starved to death. Where did bananas come from? There doesn't seem to be any other kind of plant on the planet that is related to bananas. Tomatoes and potatoes are so closely related that both can grow on the same vine. Tomatoes grown on potato vines are called pomatoes.

MORE EVIDENCE OF GODS ON EARTH

In a 1934 farm magazine there is an article about the huge grain that was found inside one of the pyramids in Egypt. This wheat was eighteen times larger than the wheat we grow in the United States. At the time they were trying to see if it would grow so that they could cross bread it with the wheat grown in America. Such a discovery would be capable of feeding the world without increasing the amount of land under cultivation.

Someone once said, "We live in a galaxy of billions and billions of stars in a space filled with billions and billions of galaxies. How can we be so arrogant as to think we are the only intelligent life in the cosmos?" Several of our prominent scientists are saying that there could be as many as one million intelligent races in our galaxy alone and there are billions of galaxies in the universe.

THE ARYAN CONNECTION

From a mythological perspective, the difference between the catholic and Catharic Christianity is perfectly encoded in the difference between the word elements ol and ar or ari that differentiate these two words.

Ol means all. All is the whole. It speaks to the desire of the Cath-ol-ic Church to be the universal religion.

Ar means before, Ari is the root for Arian, the Gnostic Christian doctrine voted out during the pivotal first Conucil of Nicea, convened in 325 by Constantine I to formulate Christian doctrine. Arians said Jesus stood between God and man. The Niceans said he was God. They won more votes.

The Arian view became heresy.

Aria al is Aryan, meaning noble, cultured, high born or pure; hence the ar element of the word Catharic.

These definitions powerfully illustrate the profound interest Adolph Hitler and Heinrich Himmler had in the Cathars. They were Aryan. This does not mean they were blond haired blue-eyed beauties. It meant they carried the noble wise blood of the god race. More importantly, they knew how to create it within themselves.

Hitler sought to create a super race through eugenics. The Cathars blood secrets were as pure gold to him.

According to Budges Egyptian Hieroglyphic Dictionary, ar means to make do, to create, to form, to fashion. It is the root of Ari, the creative god. I Egyptian, Ari-en means made by, produced by the lady of the house.

These definitions shed light on the Cathars and their prominent lady, Mari their pure Goddess.

CHAPTER SEVEN
ALTERNATIVE-3

Many believe Ronald Reagan's speech about the Space Defense Initiative, was actually a reference to Alternative-3. This special project was using a portion of its black budget to build a space based weapon system to be deployed not against Soviet nuclear missiles, but instead, to a perceived threat against hostile extraterrestrial forces.

The goal of the Alternative-3 and 4 projects was to establish manned bases on the Moon and Mars for the purpose of preserving a few people to re-seed the earth. To understand how they may have actually accomplished this feat without anyone knowing about it and doing it on a reduced budget read my book, Spaceships Of the Gods.

ALTERNATIVE -3

The 1978 'fictionalized' television documentary titled; "Alternative 3 Scenario" was shown on British television. A paperback published in huge volumes for the mass market followed this. UFO and conspiracy groups began a cult following. The topic was the theory that the Russians and Americans were working together ever since WWII. The general idea was that the Cold War is a hoax and the British, French, Americans, and Russians are in possession of flying saucers and flying saucer technology from captured German scientists. The topic is classified above top secret. One thing is clear and that is the governments of the world have been lying to us and have lied to us about certain events since WWII. Its gotten so that the American citizen and the news media wouldn't know the truth if it jumped up and bit them.

Beat writer Jack Kerouac appeared on the Jack Par's Tonight Show in 1960 with a stating: "The Cold War between the Soviet Union and the Unites States was a Hoax." Thirty years later Journalists were still laughing at jack Kerouac and his foolish statement. After the Collapse of the Soviet Union, the "Evil Empire" as President Reagan named it, decided that it had to be stopped at any cost no matter how much money we gave to the Military, is now our ally."

Monday June 20, 1977 at 9:00 PM, Anglia television based in Norwich United Kingdom aired a one-hour live TV special titled, AAlternative B3. The program was simultaneously transmitted to many other countries including, Australia, New Zealand, Canada, Iceland, Norway, Sweden, Finland, Greece, and Yugoslavia. Alternative-3 was designed to shock the nation and stretch the credibility of the viewers. At 10: PM the calls started coming in. People were told not to panic and that the program was merely an April Fools hoax that had been pre-empted until June 30. However many people were convinced that it was real. The next day the Daily Express newspaper aired a story confirming that the program was a hoax.

A respectable team of reporters who had a regular weekly program titled Science Report put the Alternative 3 TV special together. It was an intelligent documentary series, which reported on new scientific inventions. It was because of the credible people involved with the project that most people believed it was real. Why would a reputable group of scientists and science reporter's lie about something like this?

Apparently the Alternative 3 TV program hit quite close to home because the British government stepped in and stopped all future broadcasts. All people connected with the broadcast were told to deny that it had any basis in fact.

Then in 1976 'The Brain Drain' started. The best surgeons and scientists in the United Kingdom were being drawn to the United States by big paychecks. Many suicides among the scientific community were reported and other technical people just disappeared from the face of the Earth.

The real moon bases were not installed until after Apollo 11's historic landing in 1969. Before that time primitive remote landing devices containing construction materials and supplies were sent to the moon.

The Apollo missions to the moon were mostly a public display of America's ability to reach the moon. The other reason for building such tremendously large booster rockets was to haul men and supplies to construct moon bases. The Apollo astronauts were not to build the first moon colony however they did leave behind the electric jeep vehicle, which was used to transport materials for those who would follow. The men who actually built the moon colony were anonymous workers who volunteered for the job.

Both the Soviets and the United States worked together in harmony ferrying men and materials into orbit. Upon reaching orbit a ferry craft took them the rest of the way to the moon. My sources say that the first moon base was completed in 1971 and the rest were finished by 1976.

At the same time small groups sprang up such as, "Friends Of The Earth" who were just beginning to understand global warming. Then in 1978 with the help of Sphere Books Leslie Watkins, David Ambrose and Christopher Miles published Alternative 3. Ambrose and Miles didn't help write the book, but their names were included for copyright purposes because they had written the television program.

"The TV program caused a tremendous uproar because the public refused to believe it was fiction," Mills said. "Letters poured in from all parts of the world, many from highly places people in position to know what was going on. This convinced me that I had accidentally trespassed into an area of top-secret truths. To sum up the book was fiction based on fact. I had inadvertently got very close to the truth."

President Reagan said in a speech to the United

Nations General Assembly September 21 1987 about the need to turn swords into ploughshares. Reagan said: "In our obsession with antagonisms of the moment, we often forget how much unites all the members of humanity. Perhaps we need some outside, universal threat to recognize this common bond. I occasionally think how quickly our differences would vanish if we were facing an alien threat from outside this world."

February 16, 1987 Gorbachev announced at the Grand Kremlin Palace at our meeting in Geneva, "The US President said that if the Earth faced an invasion by extraterrestrials, the United States and the Soviet Union would join forces."

NASA inadvertently provided documentary evidence of a Star Wars defense system being deployed against UFOs traversing just above the atmosphere when it released video camera footage of a UFO suddenly making a 90-degree turn to the right and accelerating off into space a split second before a burst of light and a streaked high-energy pulse shot up from earth where the UFO used to be.

Why are they shooting at extraterrestrials? Richard Hoagland, former editor of Star & Sky magazine explains: "There is a great deal of naiveté' about forces in our society that do not want this knowledge, even in the form of a very useful technology, to get out." "There are interlocking institutions whose job it is to keep themselves in business. And they do not take lightly to this ET technology. After all, how do people stay in power? Because they define the universe as limited, rare and scarce they then can put a price tag on it? Then they put people in charge of doling out the scarce precious thing; other people buy it at whatever the market will bear. And you have a really neat system for controlling people."

It was known as far back as 1950 that extraterrestrial were living on the moon. A veteran who tells the following amazing story contacted Allan Carmack doing research for the TV show Strange World.

"Thirty years ago, my Supervising Sergeant called me aside and informed me that there was a technical problem in a highly-classified area elsewhere on the air Base. At the time, all systems were expanding to support increased military efforts for the Viet Nam War. As a part of that, it was my job to support and maintain highly classified systems recently installed in a new Top Secret facility on the Base. Our unit was under the command of the director of Intelligence at headquarters SAC, Tactical Air Command, Langley Field, Virginia.

"My supervisor stated that the Lunar Orbiter Program had encountered a problem with an Electronic Photographic contact Printer, identical to equipment that was utilized in the darkrooms of our own Unit. This was the first Lunar Orbiter Program, the purpose of which was to bring back the first close-up pictures of the surface of the moon. These photos would later be utilized to select an appropriate landing site for the first manned landing on the moon, in 1969.

"As the only Electronics Repairman on the Base with a Crytplogical Security Clearance, a step above Top Secret, I was being loaned to the project to see if I could resolve the system problem. More than excited at the prospect of helping out and possibly having a chance to view the first close up photos of the surface of the moon, I was briefed on security and gathered the appropriate equipment and tools for the task.

"Driving across the Base on the perimeter access road that skirted the flat dusty fields and long runways in the distance, I noticed an experimental helicopter hovering fly-like in the air just above and to the south of the massive arching metal-gray hanger, one of the largest on the base that housed the Lunar Orbiter project."

"Upon entering the hanger, I was asked to present my Top Secret Identity Badge. In exchange for their internal higher-level identity badge, this was to be worn around the neck on a chain. Another guard escorted me through a series of security doors to an expansive open area within the hangar."

"A large black fabric curtains hung from a metal grid suspended from the ceiling. These, in effect, cordoned off various working areas within the larger space. Passing through on of the draped areas, I entered a large open space where people in small groups stood talking quietly, with a sense of seriousness and concern."

"I was immediately struck by the number of people who were present, who appeared to be civilians, and also some scientists from other countries. With a bit of instant shock and judgment, I thought to myself, why are they here? I had a very strange feeling, a feeling that something is off here, something is not quite right."

"I was introduced to a man dressed in civilian cloths and a lab coat, the head of the project, a Dr. Collie, I believe. In a very gracious and reserved manner, bringing to mind an image of Sherlock Holmes, he softly conveyed to me that the equipment in question was holding up the processing of the first close-up photographs of the surface of the moon and also delaying the release of photos to be provided for study and release to the world, and how grateful the program staff would be if there was anything that I could do."

"An Airman escorted me into a darkroom. Inside another young Airman assembled strips of high-resolution 35 mm film into what is called a mosaic. He was placing side-by-side successively numbered photographs scans of the lunar surface, which had been transmitted beck to earth from the Lunar Orbiter. Each surface scan covered a narrow band of terrain, and successive orbits around the moon were required to assemble a complete photographic image of the lunar terrain."

"Left alone in the faint red light of the darkroom with the Airman and equipment, much of which I had never seen before, I began to question the technician, attempting to discern what the problem might be with the ailing contact printer. After a few minutes of investigation, it was clear there was a problem with the electronic control circuitry, which was comprised of several small plug-in modules.

"Having no spare parts on hand, it was clear I was going to have to trouble-shoot each module on a component-by-component basis, a very tedious and time consuming process at best. This was something that could not be done in the faint red light of a darkroom. The unit would have to be removed from the darkroom and taken into a more appropriate space to allow for the accomplishment of the task.

"Talking with the Airman on the other side of the room, questions floated into my head. I was curious and fascinated with the whole process. How were the signals from the lunar Orbiter transmitted to the lab? How were they converted into images on photographic film? How were the images corrected and aligned into the final mosaic negative?

"I knew there were questions that I should not ask, in fact I was alone with an Airman who was obviously as enthusiastic as I was about his job. I hesitated for a moment, weighing in my mind what I should do.

"Under normal operating conditions, many other people would have been in the lab; part of the assembly line of production. But, here we were all alone, so I began to ask all those questions.

"After about thirty minutes of technical discussion and a complete rundown on all the steps in the process, the Airman turned to me and said candidly, "You know they've discovered a base on the back side of the moon!" I said, "What do you mean?" and again he said, "They have discovered a base on the moon!" and, surreptitiously, at the same time dropped a photograph in front of me.

"There was a mosaic print of the surface of the moon, with some sort of geometric structures clearly visible. Scrutinizing the image, I could see spheres and towers. My first thought was "Whose base is it?" Then I realized the full implication: it was not anyone of this earth.

That was more than thirty years ago. Since then NASA had photographed many more structures on Mars, Venus and the Moon. Don't you think it is about time mankind grew up to face the awful truth: "That we developed high technology several times in the past, left the planet to colonize Venus, Mars, and the Moon and were annihilated either by a solar flair or some other cataclysmic event such as a meteor shower?" The surface of Mars, Venus, and the moon are littered with giant buildings, pyramids, bridges, domes, towers, glass tunnels, pipelines, and other ruins, which are the remains of past civilizations.

If NASA were to show these ruins to the world then there would be no need for war here on earth. They won't do it because it would put the arms dealers, governments, churches, and the military out of business. We really need to grow up, develop other sources of energy, and get off the planet before we kill ourselves by burning up all the oil and coal, which is causing global warming. We need to get going to become a type one civilization again; like we were over ten thousand-years-ago.

CHAPTER EIGHT
ARE WE ARE A SOURCE OF DNA?

Modern day abduction cases currently number in the millions. Victims describe various sampling of their blood, bone marrow, hair and skin. After the abduction victims quite often describe small depressions in their skin or scoop marks made by some kind of small metal scoop. It's as if their society is so old that their DNA clock is running down and they need to replenish it.

Then there are the cattle mutilations. It's as if a razor were used to remove the testis and other vital organs for future study. Scientific examination of the serrated cut reveal that whatever they used to make the cut has the ability to cut between the living cells. We no technology like that! It would take several hundred years of scientific development for us to be able to manufacture such a device.

I addition, every dead animal mutilated in this was completely drained of all traces of blood and there was no trace of it on the ground. Where did the blood go? Apparently they were taken up into some sort of craft completely drained of blood and then dropped from a height of ten to twenty feet.

I don't believe that cattle mutilations have anything to do with the inter-dimensional ships. They are the work of the 'Grays' or some other less advanced race that consumes blood. One of the things you don't hear in the news is the fact that approximately two thousand people have likewise been found completely drained of blood with various parts of their body missing and it appears that the body parts were removed while the victims were still alive.

BOEING

In other news, Boeing's new TRW flying laser cannon disappeared sometime in mid December. Nobody can find it. The FBI and other government people are snooping aground Puget Sound. Phone calls about this information are being cut off. Lets hope our military got it and not the North Koreans or some terrorist group.

Boeing was experimenting with a high ultraviolet frequency, which is similar to what a neutron bomb puts out. It would have been the ultimate death ray weapon but it is now gone. Such a cannon would also be a threat to alien craft, which can be seen by video cameras filming in the high ultraviolet. We don't know who has this death ray. I believe it is capable of taking out Saddam Hussein or anybody else for that matter from the air. All the information on Boeing's web site has been replaced by old technology. Did this advance weapon system pose a threat to the inter-dimensional ships overhead?

END OF THE WORLD

Many of the world's leading scientists attended a 1987 meeting at Lawrence Livermore Laboratory; the topic was The End of the World.

Asteroids capable of wiping out one quarter of the world's population strike earth on average once in every million years. Smaller bodies capable of wiping out a major city hit once in every two or three hundred years. The general consensus of the meeting was that impacts put the entire planet at risk.

The answer to preserve humanity is we must advance our technology to the point where a few of us can live for extended periods of time in a higher dimension orbiting earth, possibly merge with or join the angels already living there.

As astronomers gaze out into space in all directions taking the pulse of space so to speak they notice that about once every week there is a large explosion of such violence that it can only be a star going supernova. These big bang events usually are followed by an intense burst of cosmic radiation signifying that it was a star that exploded. These are natural events for stars. When they reach a critical point of mass or burn up all their energy they turn into neutron stars, black dwarfs or black holes.

About one in ten of these explosions doesn't give off cosmic radiation. There is no explanation for these events. The general consensus among scientists is that developing civilizations playing around with fundamental building blocks of the Universe may be causing these explosions. Their atom smashers trigger an event so massive that it wipes out whole stars and solar systems within a radius of sixty-five light years. Of course if such a thing is really happening there would be no evidence left to make this determination. However, It does make one stop and wonder if the scientists in white coats who are playing around with various kinds of tokamaks, cyclotrons, and atom smashers really know what they are doing? I believe it was this doubt that made that triggered congress to cut the funding for the big one being built in Texas. It was so big in fact that it would have created mountains of data larger than the library of congress every week. An underground ring fifty miles across costing fifty billion dollars. It took three billion dollars to dismantle the thing and fill in the hole.

The latest information is that four percent of the Universe is composed of what we know as three dimensional matter. Twenty-three percent of the Universe is composed of dark matter and seventy-three percent is composed of something we know nothing about but our scientists are calling it dark energy.

They are pretty sure that supernovas create black holes. Now they have discovered hypernovas. Hypernovas are so powerful that they can destroy two thirds of an entire Galaxy. The closest star system containing a red giant star, which could go super or hypernoava is Betelgeux (Alpha Orionis, bright red super-giant in the constellation Orion). However, more than likely if it does explode it will only produce a strong gamma ray burst.

CHAPTER NINE

DOOMSDAY SHIPS

YOU OWN THEM. YOU PAID FOR THEM. THEY ARE THE BIGGEST SHIPS IN THE US NAVY BUILT AS NOAH'S ARKS FOR BUREAUCRATS. NO TAXPAYERS WILL BE ALLOWED TO GO ABOARD THESE SHIPS EVEN THOUGH YOU PAID FOR THEM.

When I first learned that two Navy doomsday ships backed up against the side of a hill near Tacoma, Washington were real I couldn't believe it. They are hidden from view and are being guarded by a submerged submarine twenty-hours a day. The ships are the largest ships in the Navy. They have no markings on them except for a number and are not found on any list of ships. They are constructed of heavily armored steel plated and they have cranes to lift boats and other cargo on and off. There are no windows and they are heavily shielded against radiation. Two more of these ships are located in the San Diego harbor and there are two more of them in Los Angeles. I suppose there are more of them in every major port in America.

There is a vast number of sick, One World Government, psycho, Hitlerian types who think that there are too many people on earth. Their agenda is to kill off most of us so that they can mold the world the way they think it should be. Where's the love?

The doomsday ships were never meant to house taxpayers. These ships were made to house government personnel in case of a nuclear first strike against America. They are a sort of Noah's Ark for bureaucrats. Nuclear weapons have become much more accurate over the years thanks to advanced guidance computers and technology leaks to North Korea and Red China they are more capable of making so-called "surgical strikes." We don't want the enemy to miss and kill innocent people living in the surrounding area therefore I have decided to include their GPS (geographic position satellites) coordinates after all they are yours. You paid for them.

I thought about giving out the GPS coordinates of the doomsday ships for a long time because I considered it to be a treasonous act. Enough people already know where they are located so that if there is a nuclear attack the ships will be targeted anyway. I am pretty sure there are enough disgruntled citizens out there who know they are going to get cooked and who will gladly help the enemy. Where's the love and compassion? My family lives in the area along with many other friends. For this is reason I have decided to include the GPS readings in this book after all.

THE FOLLOWING IS A REPORT FROM A CONCERNED RESIDENT:
Reserve Merchant Fleet or
DOOM'S DAY survival ships?

Reserve Merchant Fleet, at least that's what the neighborhood was being informed of, as well as for their discovering the $5,000 per day payment to the county and/or city of Tacoma, for the use of the essentially undeveloped (nearly derelict property) secluded deep water dockage space.

The hillside residential neighborhood individual by which I discussed the two ships, doesn't recall ever not seeing those mammoth ships moored and, I personally can't recall not seeing them for at least ten years. You can do the math, that's a serious metric ton of moneys and, that's not even including shore-power (Metered 2,400 Amps Shore Power of at least 480 V 3 phase, even a conservative partial load is another $2,000 per day), other utilities and contracted services not included comes to $25,550,000.

If you merely focused upon the upper most pilot deck house and related superstructure of associated outfitting's without knowledge of the underlying vessel, one could not distinguish the pilot house of either of these supposedly "reserve merchant" ships as from that of a military frigate or of any number of other battle qualified Navy ship.

As to the main forward deck, this massive deck is simply not well suited for any sort of freight, there's a couple of dozen elevators like structured deck access enclosures, the forward most hatch is more likely suited to covering missile bays and/or of specialized aircraft then of cargo, whereas the aft ramp is clearly over-built to handle anything that can be transported by truck. The overall construction surpasses (by far) anything "merchant" and the exceptional size and quality of those enormous side view (nearly bay-window-like) windows are hardly of any new maritime union regulation nor of standards for that of Merchant Marine ships. The fact that these two ships have been sustained in the best of readiness for over a decade parked at $5,000+/day), clearly indicates a highly unusual nature or priority of need or of potentially secret agenda.

Because these ships are massive, I estimate that each vessel could easily accommodate 10,000 individuals (with plenty of elbowroom in addition for dozens of the largest semi tractor/trailers) and of sufficient provisions to sustain themselves for months at sea, obviously more if re-supplied at sea. The present station keeping skeletons crew necessary would have to be nearly 100 individuals per ship (obviously far more if becoming fully operational), that's not including others contracted and/ore the members involved with delivering the mega-tons of whatever they elect to store onboard.

The pictures speak for themselves and, should be considered as WYSIWYG as you'll ever get to anything that's under NSA/DOD command. The otherwise supposed "reserve merchant fleet" functionality had likely been a cover, not that such capable ships couldn't provide commercial transport, just not affordably, thus not competitively in any real world that you or I happen to live in.

Here are a few web sites associated with **"SPERRY OCEAN"**
http://www.sperryoceanterminals.com/
http://www.psmre.org/images/2001/2001images.htm

Tons of location and surrounding countryside photos but

not one image of either ship (*even though both ships were always there except for a brief week at sea*). I've tried a little web research to identify ship specifications and so far, "no cigar!"

http://www.portoftacoma.com/newsreleases.cfm?sub=68&1sub=267

"At the port, Army reserve trucks were rolling-onto the Cape Island, (*a fast Sealift ship owned by military Sealift Command*), in a mock demonstration of preparing and loading equipment for an overseas voyage to a third-world, unimproved port. The ship is one of two that are normally berthed along Tacoma's Schuster Parkway, at the Sperry Ocean dock."

NOTE: if these vessels are those classified as "fast sealift" then, due to their extremely small stack configuration (*sufficient for large generators*), the ships could be NUCLEAR. In the perspective views of either ship there's insufficient stack capacity for conventional propulsion systems. To the eye these items seem woefully inadequate for the type of energy required to qualify these vessels as "fast sealift," unless by fast they mean possibly 16 knots (*down wind*).

COMMERCE BUSINESS DAILY ISSUE OF SEPTEMBER 30, 1996 PSA 31690

"Department of Transportation, Maritime Administration, Office of Acquisition, MAR-383, 400 Seventh Street, SW. Room 7310, Washington, DC 20590"

http://www.fbodaily.com/cbd/archive/1996/
09(September)/30-Sep-1996/Vawd001.htm
"Clayberthing Services for two ready reserve force vessels
POC Ben Burnowski, (202)366-1932. CNT DTMA91-
96-D-00004/DTMA91-96-B-00007, AMT Estimated
$4,701,022. DTD 091796. To: Sperry Ocean Dock, Ltd.,
2201 N 30th Street, Suite D Tacoma, Washington 98403-
3320. (267)"

"The pictures I've taken were from public property
sites and as such not as much detail as there is to be seen
in person. A better camera and more effort at framing
could produce a better perspective. If these ships remain
at moorage, I should be able to acquire a couple of better
shots. From what I read, There are at least two more of
these ships, as big or bigger."

CHAPTER TEN

MAGIC SCIENCE OF THE FUTURE

Is a book by Joseph F. Goodavage published 1974 by Signet Classics, Mentor, Plume and Meridian Books. Mr. Goodavage has written and published dozens of articles, both for scientific and general audience publication. Formerly science reporter and writer for the New York Daily News syndicate, The Chicago Tribune and the New York Newspaper Guild.

"It may be that Homo sapiens isn't all that unique in the cosmos. Perhaps there are great civilizations spread across the Galaxy (and beyond) composed of undreamed-of variations of the genus homo. While we discover more about the human family and the riddle of evolution and as the mystery surrounding man's true origins are increasingly complicated, the clearer it becomes that other species of (quite possibly) non human intelligent, rational beings have evolved equally complex brains and languages. There are undoubtedly great alien cultures of non-human civilizations somewhere in space. If this as likely as it seems then it's even more likely that representatives of those civilizations could have visited Earth at any time (perhaps during our 'prehistoric' periods) and left artifacts or some other trace of their presence. This theory is now regarded as scientifically plausible.

"Technology stands at the crudest, most cumbersome era of space travel. Our mighty cannibalistic rocket, fueled by million of tons of costly chemicals, are the 'gas bags' of space voyaging. Compared to the fantastic vessels of the futures, today's powerful boosters, hot-air-inflated linen bag, or to the Write Brothers, stick-and-cloth prototype of the airplane."

"Ever since American astronauts began to leave traces of their presence on the moon and soviet and American space probes started parachuting sensors into the atmospheres of Venus, Mars, and Jupiter it has become easier for us to understand how the same events could, and quite possibly did happen eons ago."

"It began to look suspiciously as though we're intimately involved in a Grand Design which we're only dimly aware of. The pieces are beginning to fit however, and the clearer it seems that we are destined to explore and colonize the planets and then the stars.

Stupendous as it seems, such and adventure may be as common an event in each Galaxy as are hundreds of thousands of graduating classes on all educational levels throughout the world. If the 'schedule' is kept, we will witness the withering and gradual elimination of competition and war as tools of international diplomacy, and the appreciation of man as a unique entity in the Galaxy.

"Aside from conquering disease and increasing out knowledge, power, and the ability to bring human reproduction into balance with earth's resources, the major nations of our planet are desperately engaged in hammering out treaties to avoid nuclear holocaust that might forever extinguish all trace of man in the universe. We posses this capability, plus the freedom to choose either course. Either we accept fully the responsibilities of stewardship of our fragile, delicately balanced planet or we do not.

"Are we prepared for the final examination before graduation from the planetary grammar school?

If we are to make it, it's imperative that we gain total control over the animalistic promptings of the right brain and stifle its negative biological emotions, fear, greed, and superstitions, and begin to live in a spirit of mutual trust, cooperation, and love."

STEVEN BASSETT

In the wake of Congressman Steven Schiff's death from cancer, Steven Bassett has decided to get serious about challenging government secrecy and has announced he is running for congress in the eighth Congressional District which encompasses Montgomery County, Maryland. As many of you are aware Congressman, Steven Schiff of New Mexico announced publicly he would make the government come clean about UFOs.

At long last, someone has stepped forward to boldly run for public office on the UFO disclosure issue. This was perhaps inevitable and in a way, it is amazing that it took fifty years to happen. But until now no politician was willing to risk the ridicule. Because the media giants treated the subject tongue-in-cheek, the public typically reacted with chuckles to anyone who took it seriously, and, in politics to be laughed at is tantamount to certain death. But thanks in part to the efforts of respected researchers like Budd Hopkins and Dr. John Mask, the events of the past decade have slowly but surely made the entire subject very respectable and worthy of debate, and so not the climate is right, and Steve Bassett has announced he is running for Congress.

Like Dr. Steven Greer, Bassett has come to the task from an unrelated background. After obtaining a degree in physics, he spent 15 years in business development consulting. Then in the early nineties, he became interested in UFOs and in 1995 volunteered to work for PEER (Program for Extraordinary Experience Research); the organization founded by Dr. John Mack in Cambridge, Massachusetts to research and investigate the alien abduction phenomenon. This was the spin-off from the personal research Dr. Mack conducted in his Harvard psychiatric practice that resulted in the now famous and highly controversial book, *Abduction*. Evidently this experience convinced Bassett of the UFO reality, and set him upon a crusade to assist all those who were laboring to get the information out to the public.

After five months at PEER, Bassett left Cambridge in July 1996 to set up a consulting practice to provide professional support to UFO researchers and to lobby at the national level on behalf of UFO/ET research/activist organizations. His company, Paradigm Research Group, in Bethesda, Maryland also does media liaison, funding

proposals, and secondary research.
www.disclosure2003.com.

     Living in the higher dimensions seems a natural evolutionary goal given the various cataclysmic events that impact earth from time to time. Such events that might periodically wipe out life on earth are nuclear war, pole shifts, meteor impacts, solar flares, and close encounters of large gravitational objects that cause massive tidal waves thousands of feet high. Living in the upper atmosphere in a higher dimension would allow beings like us to escape such disasters. Such a life is not without its problems and complications.

     What if a person could really live 300,000 years? Think of the long-range goals you could have. You could do the real big projects like genetic manipulation of a species or terra forming. A person could learn every foreign language on earth, or play every musical instrument known to man and become knowledgeable about any subject. Living a long time would tend to lend more meaning to life but it poses a whole different sort of problems. Your car and house would wear out many times before you reached the point where you would no longer need these items.

     "What we visualize tends to come true. The Universe is set up this way. It is very important that we visualize a positive future."

CHAPTER ELEVEN
GLOBAL WARMING

Currently we are faced with a problem of global warming. I used to think that there was no threat to global warming and if there were the changes would be so slow as to be practically unnoticeable. I believe all the rhetoric that erupting volcanoes did much more damage than humans dumping more pollution and carbon dioxide into the atmosphere than all of human habitation combined. Recently the tide levels where I live are a foot higher than usual. I am starting to take notice. Maybe there is something to this global warming fad after all.

Nikola Tesla wrote in his personal journals in the 1920's that he thought the buildup of gases caused by manmade and natural pollutants would eventually cause the planet's overall temperature to increase causing the polar ice caps to melt. As well weather patterns would change bringing fierce storms, flooding, and draughts in some areas. "There would come a time when agriculture would be destroyed by burning heat and flooding. Mankind would starve and the atmosphere would dissipate."

Since World War II the amount of carbon dioxide in the atmosphere has increased twenty-five percent. I believe burning things causes this. So far we have burned up half the oil reserves that took billions of years to make. The two greatest oxygen producers on the planet are half gone. So fare we have slashed and burned half the rain forest. Half of the phyto (plant) plankton in the oceans has been killed off by oil spills and changing weather patterns.

Satellite photographs of the Polar Regions clearly show that the ice caps are melting. Currently you have over ten thousands of people living in Antarctica. Each person consumes at least two hundred gallons of diesel fuel per year. That doesn't take into account the fuel needed to transport the fuel. Then they have these tremendously large boilers that burn several thousands of gallons of diesel fuel per hour to melt caverns and holes in the ice. They use huge tractors to drag all this equipment around that burn a hundred gallons of diesel oil per mile. Add to this all the pollution and soot buildup from this year-round human activity and you have a recipe for disaster. Believe it or not this is all being done in the name of science to study the Antarctic. Whatever we study we destroy.

On top of all the scientific activity you have thousands of tourists going down year round. I hear that tourists are even going to Antarctica in the winter. What next?

MORE RECENT INFORMATION

Scientists are now thinking that the buildup of carbon dioxide is creating more carbolic acid, which is increasing the acidity of the oceans. This may be the reason why all the choral reefs all over the world have stopped growing. It may also be the reason why about half of all the world's plankton has disappeared. Kill off half the plankton and you will kill off over half the fish. When do we tell the oil cartels to stop producing oil and switch to hydrogen. Do we have to wait until all the fish in the sea are gone people are starving?

Tesla theorized that the planet's pollution problems were being closely monitored by extraterrestrials that were studying Earth and its inhabitants. He was undecided weather or not their intentions were hostile, friendly, or indifferent. Tesla wrote that, "If it were not for the unusual way I am getting this knowledge I would have dismissed it long ago as the ravings of a mad man."

Ronald Reagan's Space Defense Initiative speech was attributed by some as a branch of Alternative 3. This special project was using a portion of its black budget to build a space based weapon system to be deployed not against Soviet nuclear missiles, but instead, to a perceived threat against hostile extraterrestrial forces.

Although many of Tesla's ideas were scoffed at during his lifetime they are now being sought after by the military and many governments around the world. Tesla's ideas of electric death rays and particle beam weapons are now being used in space. UFO's have always been around as evidenced by various passages in the Bible. After the beginning of the 20th century their numbers increased greatly. It's as if we have suddenly become more interesting to them. Our military on the other hand perceived them to be a threat because of the blatant way they could fly around and they couldn't do anything about it. This is one of the reasons why the military must deny their existence. If they admitted that UFOs actually existed then they would have to admit that they couldn't do their job of protecting us from them. The very definition of a bureaucrat is, "They can never admit a wrong." They will never admit that they cannot do the job they are assigned to do no matter how incompetent they may appear. When a UFO appears overhead it puts them in a catch 22 situation. They have to deny it exists, and call it swamp gas.

CHAPTER TWELVE
Too Many People On EARTH

Our government and the news media are constantly telling us that there are too many people on Earth. It is a lie. If we didn't waste our natural resources Earth could sustain a population of 11 billion people. It seems that the rich who control government policy so that they can get richer are the ones behind all this. How did such a disconnect occur? How was it possible for this to happen in a democracy? When and why did the government and the people become adversaries? Is it just a question of citizen apathy, or does the government take cynical advantage of this apathy? Is the mass media a willing ally in a form of mass conditioning or mind control?

Three modern phrases have come to represent to an unacceptable degree the relationship between citizens and their government. They are, A "Need to Know Basis", "Don't Ask, Don't Tell", and "You can't Handle the Truth." These concepts are often stated verbatim in actual policy and reside at the center of an increasing patronizing posture-the government acting *in loco parents*.

A flagrant example of this disconnect is the ECHELON surveillance system being operated in the U.S. by the super-secret agency, the USA. ECHELON is an automated global communications interception and relay system operated jointly by the intelligence agencies in the U.S., and the United Kingdom, Canada, Australia, and New-Zealand. This highly sophisticated satellite-based system may intercept as many as 3 billion communications every day, including phone calls, e-mail messages, Internet downloads and satellite communications. It is estimated the ECHELON sifts through about 90 percent of all traffic that flows through the Internet, and yet most Americans have never even heard of it.

Spying/surveillance technology has outrun Congressional oversight, and it is time for the citizenry to get into the act to comprehend exactly how Big Brother actually operates, and how it may be abusing it's powers.

Another example of government intrusiveness is the IRS tax code. Using the data that you are forced to send to the IRS each year it can profile and determine how much disposable income you have, what you are spending it on and when and where you will be at any given moment. We definitely need to change the currently illegal IRS tax code to a legal flat tax based on your income. It is none of the government's business where and when you spend your money.

There are currently a little over six billion people on this planet. Now a planet with 11 billion spiritually awakened, people cannot be controlled. Such a scenario would change the harmonic frequency of the whole galaxy. The vibration would be love, and the bad guys don't like that.

In 30 years when the population reaches 10 billion we will be consuming almost everything the planet produces. We cannot live beyond our means. We are losing a forest area the size of Great Britain annually. Twenty percent of all species of plants and animals are gone forever and in forty years at the present rate of extinction we will have lost fifty percent of all plants and animals on earth. We cannot continue to live beyond our means.

Our leaders are so corrupt that their systems are breaking down. Instead of acting responsibly, by releasing control and letting people take more responsibility, the corrupt leaders want to kill off most of the people in an attempt to maintain the status quo. There is a way of speeding up the spiritual awaking and that is what this book is about.

HOW TO PROTECT YOUR FAMILY FROM FALLOUT

The theory for survival aboard ship during an atomic blast is outlined in my book: **The Big Score**, which may be browsed on the Internet at:
www.mittymax.com

During the Bikini Atoll test in the South Pacific a Japanese ship actually sailed under the mushroom cloud. Not one of the crew died from the radiation. A few of them did have some radiation burns and most of them got sick but none of them actually died. If they had taken proper precautions they wouldn't have had any kind of radiation sickness. Most fallout particles are very fine; the size of an atom and it is quite difficult to filter them out of the air unless you have a filter small enough to filter out atoms. It is just about impossible except for one thing, carbon wool.

When radioactive fallout from a nuclear bomb comes down it accumulates on the roof of your house. As rain washes it off the roof of your house it piles up on the ground next to the wall of your house. If you are hiding in the basement you are going to get the biggest dose of radiation. If you stay in the attic and wash the roof off with a hose once in a while the radiation in the attic won't be near as high as at ground level. Wherever radioactive dust accumulates the radiation levels build up becoming stronger and stronger and eventually you get sick and die from it.

If you are aboard a ship you can send someone outside wearing rain gear every four hours at first to wash the deck off with a hose. The radioactive dust is washed overboard and doesn't have a chance to build up to high levels. The person going outside is exposed to the radiation for only a few minutes thereby limiting his dosage. If you rotate crewmembers for this duty their individual dosage will be limited to acceptable levels. The rain gear protects him from getting the dust on his skin and if he wears a gas mask packed with carbon wool it will limit the amount of dust getting into his lungs.

In the case of the doomsday ships I presume they have a way of flushing the dust off the decks with water automatically so that the people don't have to go outside. If the ship itself is heavily shielded the people on board might even survive a neutron bomb attack. Neutron bombs are the current weapon of choice because they make bomb shelters obsolete and leave behind very little radioactivity. The neutrons pass right through most shielding killing all living things.

If the Red Chinese were to attack America with neutron bombs and other atomic weapons it will kill off practically every living thing. The government plans to put some of their people on doomsday ships to start over again with the "New World Order". In the meantime our atomic submarines and land-based missiles will wipe the slate clean on the other side of the globe. I wonder if they plan to take any goats and sheep aboard these ships so that the people can have food or maybe they plan to clone livestock from frozen embryos when the radiation levels become acceptable. I can't imagine living on C-rations for five years.

Space is a dangerous place. There is always the possibility of some kind of cataclysmic event that could take place; exterminating most of us on the planet in one fell swoop! Maybe it is a good idea to have a few doomsday ships to re-seed the earth. The tenth planet Nibiru, that Zecharia Sitchin spoke about in a lecture is inbound and is now visible in the southern sky with a small telescope. It will pass between Mars and Jupiter some time in March 13th of this year 2003. Nibiru is a large reddish-brown planet four times the size of earth and it may be composed mostly of ice. People who study the sky say it is already visible with a pair of binoculars.

Besides the doomsday ships the government also has tunnels running through solid rock all across America that were cut by a railroad tunneling machines. It uses flame and a big rotary drill bit. These machines can drill up to five kilometers through solid rock. They have tunnels extend from Silicon Valley to area 51, Black Mountain Colorado, Utah, New Mexico, north to Washington DC, and a lot of other places I don't know about. If anyone tries to invade this country it will take them a very long time to subdue the survivors. It definitely won't be worth the effort but then when you are dealing with insane extremists anything is possible.

If there is ever a nuclear attack you must have some potassium iodine or sodium iodine tablets on hand to give to your family. Even taking a few drops of iodine in a glass of water every day is a good idea. The theory behind this is by taking iodine it saturates your thyroid with good iodine so that it will not absorb radioactive iodine. You want to eliminate any chance of a build up of radioactive material in your brain to avoid brain cancer.

WILLIAM ABBOT

When William Abbot refused to develop biological and chemical agents for the Korean and Viet Nam War the government discredited him. He couldn't get a job anywhere. Trans World Airlines learned of his mathematics capability and hired him as a navigator for their Pacific

Ocean flights to Hong Kong. William would sit in the navigator's chair and read pocket books. When the pilot asked for a fix he'd take a couple readings with a sextant out that little window in the ceiling above the navigator's desk, jot down a few numbers on a piece of paper and hand it to the pilot. William was such a brilliant mathematician that he could do calculations in his head that usually took a professional navigator twenty minutes. He also worked for Pacific Northern Airlines. The government ruined William Abbot's career and Union Carbide stole his patents causing him to die of a coronary heart attack at the young age of 48. He left behind a wife and four kids. He might have gone on to make many more discoveries that could have benefited mankind.

Several years after William Abbot passed away his daughter Gretchen was invited by a neighbor and fellow scientist who also homesteaded part of Yukon Island in Alaska to inspect a Quonset hut filled with boxes of papers and documents. Inside she found a complete copy of the papers Stanley Mott and General Funkhauser had smuggled out of Germany at great risk from Hitler's Secret rocket base, Penemundy.

Thanks to William F. Abbot the inventor of carbon wool in the 1950s the government has filters fine enough to catch atomic dust.

I met Mr. Abbot and knew his family. William Abbot had a laboratory set up on Yukon Island in Kachemak Bay, Alaska. The development of carbon wool was collaboration between Mr. Abbot and a German scientist Freidrich.

Carbon Wool with its exceptional heat insulating properties can withstand thousands of degrees and was used in the nose cones of rockets.

WHAT WE MUST DO

According to American Indian legend we are living in the fifth world. This means that there at least four other advanced civilizations that came before us that lived twelve, twenty-five, and fifty thousand years ago and for one reason or another were annihilated either by natural causes or self inflicted genocide. Several of these civilizations made it off the planet and built structures on Mars, Venus, and the Moon. Archeologists are finding layers or radioactive material below the bones of Neanderthal man and below that they are finding the remnants of ancient civilizations. Obviously after the events that wiped the slate clean mankind degenerated to the level of the cave man who chipped rocks and hunted with bows and sticks to survive and then was brought back into the light of self realization and higher enlightenment by outside forces.

The ancient Egyptians and Sumerians knew practically all of our knowledge of philosophy, plumbing, architecture, mathematics, language, astronomy, and science. The Gods also gave this knowledge to the early Greeks and Romans and some of it was passed down to us through their ancient writings. It's too bad we destroyed most of our early knowledge. The most ancient civilizations possibly received their knowledge from, (if we are to believe the legends) from the Gods and other civilizations that existed in prehistory like Atlantis.

Another scenario that is becoming increasingly more evident by the numbers of artifacts that are being found on the planet's surface is that modern man originally came from the planet Mars about three million years ago before the pole shift and massive meteor impact that decimated the planet causing it to loose it's oceans. I believe that mankind's seeds probably came from some place else, from some other planetary system and his real origins are perhaps billions of years old. In Illinois they have found complete 'human' skeletons buried in layers of coal that are over 160 million years old. This kind of puts the KIBOSH on the theory of evolution.

Because most of this ancient technology worked with nature and was not destructive to the environment I believe some of it was given to mankind from other alien races who through millions of years of trial and error evolution were able to achieve a type one civilization which went out into space for the purpose of assisting other developing life forms in other biospheres. (A type one civilization is one, which does not burn fossil fuels. It also does not kill his brothers and sisters through tribal warfare.) Most of the ancient technology did not evolve on earth but came to earth from other civilizations that made the same mistakes we are making now. Some of these ancient civilizations were able to overcome their problems and came to help us, which is what good neighbors do.

Undoubtedly evolution was helped along from time to time as gifts were passed down from above. Practically all of our foods such as the tomato, potato, and corn were genetically engineered to have a lower alkaline content for mankind's benefit. Adam and Eve were made aware so we could shoulder our own responsibilities. We must do our part as the book says or we won't make it as a people. Instead of being what we think is the smartest creature on the planet we could turn out to be the dumbest.

Michael Cremo writes about complete modern human skeletons (as opposed to Neanderthals) found under rock strata in Illinois that was more than 200 million years old.

I have no doubt that humans developed high civilization several times in the past between ice ages. We have ten to twelve thousand years between ice ages to get our act together and get off the planet before the next one. Check out my book Cosmological Ice Ages and Global Warming on www.alaskapublishing.com.

THE FUTURE

We must pray for our families, friends and neighbors and help them any way we can.A person can fill library with books on what is wrong with our world but that is not my objective here. I merely want to make people aware of a small portion of the most deadly developments that threaten our existence. It is very important that we keep a positive mental attitude.

SPEEDING UP THE SPIRITUAL AWAKENING

"An army that caries the Ark of the Covenant before it is invincible."

It must have seemed like magic to the ancient Hebrews when their priests would generate an electric ark (ball lightening or shikanna glory) between the outstretched wings of the golden cherub on lid of the Ark. To do it by using thought power was even more miraculous and mystifying.

There are several reports of the Ark being carried into battle. It was customary to have four people carry the Ark on long poles slipped through the brass rings attached to the side of the gold and wood box. When out in the open one could not touch it for fear of being electrocuted because it gathered energy from the atmosphere. The MANNA, MFKZT, or ORME material inside also had thousands of amps of electricity flowing through it. Such technology may have been passed down from above.

There is no way four men could carry the Ark, let alone carry it over rough terrain without breaking the thin poles unless it was an anti-gravity device. Estimates of the weight of the three-inch-thick gold lid place it at 2,500 pounds. The total weight of the Ark would have been 2,600 to 2,800 pounds depending on what was placed inside yet, it is described as floating along, sometimes carrying the people who were transporting it.

It was customary to carry the Ark at least 800 yards or more in front of the Hebrew army. Once within range of the enemy a priest would concentrate on the manna inside the Ark and generate the "shikanna glory," commanding the opposing army to lay down their arms and go home. The commands echoed magically inside the heads of the enemy as the Ark advanced. If you can produce an electric discharge by using thought power what you are doing is amplifying your thoughts a billion times. This must have been quite terrifying for any opposing force. Those who resisted which were few; died of a sudden heart attack as their bodily functions were arrested by the powerful force emanating from the Ark.

The Ark was a powerful instrument of war yet it could be used for peace. Gold dust or nuggets placed on the brass pan or mercy seat between the outstretched wings of the cherub was transmuted into the manna at 5,000-degree temperatures. This was the powerful manna, the food of the gods.

From: *Philosopher's Stone* by Henry Kroll

Moses goldsmith, Bazaleal, made the white-powder-gold or manna using the ark on top of the Ark. It is being made today by burning gold in an argon filled chamber at temperatures of 5600 degrees. This changes the shape of the atoms from a spherical shape to a conical, mini-universe, shape. It then become the most nourishing substance on earth where a person reverts to the physical age of 26 years and can possibly live as long as the Anunnaki themselves which live 350,000 years. It feeds the sole and corrects the DNA.

The Anunnaki Gods created the 120-year life span for humans on Earth by restricting access to the bioorganic powder so that the humans wouldn't overpopulate the earth. These revelations may or may not affect the Church because people are slow to comprehend the meaning of the ancient Sumarian translations. It is true that modern humans suddenly appeared in the form of Cro-Magnon man about 300,000 years ago, with no evolutionary evidence. Archeologists are puzzled by this and cannot offer any reasonable explanation for it. Occum's Razor and Sherlock Holmes concur: "The only remaining answer no matter how implausible must be the right one."

The monotheism (belief in one god) of Bible translations was an attempt to unite mankind in service to a common religion in the hope that it would stop wars thereby allowing civilization to take root. Europe had to crawl out of the slime of the dark ages somehow.

Father Charles Moore went on to say that: "David Hudson, the man who recently re-discovered the white-powder-gold, manna set up a manufacturing plant in Santa Barbara, California. In the first videotape of his lectures that he received form David Hudson. Father Moor said that David Hudson looked like he was in his sixties on the first videotape but in the last videotape he received from him he looks like he is in his thirties. David Hudson has disappeared from public view and cannot be found anywhere.

The drawbacks to taking the stuff are you develop telekinetic and telepathic powers. You can read minds and send mental messages. Practically everyone who has taken the white-powder-gold has moved away from cities to a very remote area because they hear and take in all the thoughts and all the suffering of those around them.

The biggest obstacle is how to handle the psychological aspect. The power to bless also holds the power to curse. First we must learn to use the power before we can consider taking the stuff. How do we reach the point to trust ourselves? First you must learn to trust the voices that you hear.

The call to become a master is a tremendous decision. Once the power begins to manifest within a person he has to retain conscious thought at all times otherwise he might accidentally cause harm to another. If you are a master healer and wipe out cancer and other diseases saving thousands of lives you still may go wrong. One of the humans you saved could invent a new weapon that kills millions. Life holds no guarantees.

The discovery that the Anunnaki gods lived on Earth does not destroy religion. Religion is about the spiritual bonds that connect us to nature and one another. The very first religion on earth before the Anunnaki came here was animism. An animist is one who discovers that the universe is made out of spirit. Conscience is the stuff that the universe is made of and our conscience is only a small part of it. Shamans have come to this realization thousands of years before the gods came from outer space. In other words, God was busy long before Moses crossed the Red Sea.

The **MANNA, MSKZT** or Anunnaki (ORME) Gold can help us discern the truth. Those who take it know all truth instantly and become clairvoyant. Such latent powers might help people become aware of the holistic paradigm thereby motivating study to usher in a unifying transition for humanity. After we learn to use it safely it might be used to amplify our brain waves so that we could communicate directly with the inter-dimensional ships.

ADDENDUM

Two years after I wrote this book I am still discovering secrets of the MANNA. I was talking to a waitress about the manna when suddenly it hit me. The reason the million or so Hebrews could traverse the desert without starving to death was because they had the technology to make the manna. The Egyptian army could not follow them because they did not have this technology.

All plants take up energy from the soil by extruding alkaline in their roots. The alkaline dissolves the minerals and other nutrients. The reason you cannot go out behind your house and eat all the plants that are out there is because most of them have too high an alkaline content and you would be poisoned. The plants that we do eat have been genetically engineered by the Gods to have a lower alkaline content so that we can eat them.

This answers a few more questions such as why native plants grow better than the ones you plant. The reason is that native plants having a higher alkaline content are more efficient at taking up nutrients. The answer to the next question is, if you could eat the native plants you would be assimilating more beneficial nutrients. Do domesticated plants contain less beneficial nutrients due to their lower alkaline content? Probably so! Would we live longer if we could assimilate the nutrients directly instead of depending on plants to do it for us? Probably yes!

Moses and his brother Aaron would boil the soil in an alkaline solution for several days. They probably used the alkaline waters from the spring of Murah. After pouring the clear liquid into another container they added a little vinegar to raise the acidity and the monatomic soup and minerals became visible as a sort of milk. Taking this into one's body was partaking of God's light or God's food. Some of the Hebrews probably mixed this solution with whatever grains they had brought with them and baked them into cakes to make a life-sustaining meal.

I was reading the Egyptian Papyrus of Ani Pyramid Text, about the Pharaoh in search of enlightenment by using the Golden tear of Horus, the MFKZT, the manna. It dawned on me that it is not only talking about correcting genetic imperfections and living forever. The last passage refers to time travel!

"I am purified of all imperfections. What is it? I ascend like the golden hawk of Horus. What is it? I come by the immortals without dying. What is it? I come before my father's throne. What is it?"

The user of the golden tear of Horus can not only outlive the immortals he or she can exist before their father's throne. The Hathor-related petroglyphs of Dendra showing two glass globes supported by djed pillars or columns may not actually be glass globes but inter-dimensional Meissner fields used for time travel. Subtle differences in the drawings suggest that a field is being represented not a glass globe. The differences in the djed support columns suggest that there is a difference in the fields. Inscriptions identify the serpents as Harsomtus, the divine child of Horus and Hathor. Many lily or lotus flowers are carved into the stone temple walls. This symbol

When two magnetic Meissner fields produced by two different superconductors come in contact time and space is altered in what the experts call a perpetual quantal wave. A person standing between these fields may actually be transported to God knows where. The petroglyphs of Dendra may actually portray an Annuaki time portal in operation.

The black beach sands of Tuxedni bay naturally exist in a perpetual quantal wave generated by the monatomic double combined atoms with high-speed electrons. This material is there and yet it is not there but **sometime else**. It gives off beta radiation, especially when subjected to bright sunlight.

Trying to understand this stuff it a lifetime endeavor.

For more information and pictures of the Venetian structures check:

http://guthvenus.tripod.com. If you do a search on Venus you will get hundreds of pages. Also: http://geocities. com/bradguth

My books may be browsed **FREE** on the Internet at:

www.mittymax.com.

0081-Spaceships Of The Gods and **Philosopher's Stone** *deals with a substance we eat every day.*
0085-Saddam Hussein And The Sand Pirates
0087-The Big Score - *An Easier Way Of Life*

You may download an e-**Book** with a credit card for $4.00 US

My books are also available in paperback for $17.95 postage paid. My new web site contains a list of my books and others, plus an array of equipment and instructions for producing alternative energy sources.

On the **Internet** it's:
ALASKAPUBLISHING.COM.
TUXEDNI BAY LODGE

To reserve a furnished cabin for a week or a month at Tuxedni Bay Lodge telephone: (907) 252-1390

TUXEDNI BAY LODGE is open for business all year round. Our guests enjoy Brown-bear-viewing, Halibut fishing, Salmon fishing, and Bird-watching tours from an open skiff. Activity rates are: $50 per person for a four-hour trip. This is a good way to decide if you should make your home in the wilderness.

Winter activities consist of Photography - Gold Mining and Cross-country Skiing.

For additional information and a brochure telephone:
907-252-1390 or 907-252-1391 or write to:

Henry Kroll
PO Box 526
Kasilof, Alaska 99610

INTERVIEW WITH KEVIN 10/5/03
Soldotna, Alaska

Interviewers note: Kevin is a big powerful man of Apache Indian and Norwegian decent weighing over three hundred pounds. Typically he works on the oil platforms in the Gulf of Mexico as a welder but is currently looking for work in Alaska. I would say by looking at him that he is capable of tremendous physical strength. His forearms are as larger as most people's thighs.

I believe that I am a good judge of a man's strength because I fished king crab here in Alaska for 25 years catching several million pounds. I commercial fished for salmon more than fifty years and owned a seventy foot crab boat. Later on I purchased a 128 footer. In other words, I am an ex-sea pirate.

INTERVIEW

[I was hoping we could talk a little about some of your childhood experiences. Earlier you mentioned something about them burying stainless steel or what looked like big stainless steel canisters. Where was this?]

This was in Florida. They had a crane...you know when you were a kid everything looks huge but I remember the oak trees, the size they were up the road and that rock road was dug up and they had a crane. It looked like ah- ah some kind of everyday normal activity like you would see somebody doing. This crane was lifting up these canisters, it was all it could do. [They were big canisters!] They were quite large. I would say about... [Five hundred pounds?] Oh No! what ever was in it was making the crane grunt to get it up. I remember ah..ah.. the load that the crane had was trying to pick these canisters up...I would say they had to be at least fifteen feet high...and I would say that two people tip-to tip could go around them. But whatever was in these canisters but whatever it was made of was extremely heavy. Because they had a crane and the crane was grunting just to pick one of them up. [It must have been some valuable mineral that they wanted that was heavy that they were burying there?]

[Where was this?] It was in Arta Creek, Florida. Arta creek Florida is about...the biggest town that most people would be familiar with is Gainsville, Florida because it has a university there and its about 70 mile from Gainsville and it would be South of Gainsville. [you were about five at the time at the time and they ignored you and didn't do anything to you because you were young and they didn't think you would remember them?] I watched them for a good...for a good while...about forty minutes. [What did they look like? Did they look like your little guys?] No! They looked like normal people but this was...it wasn't what they really would look like. It was...they looked like people would think normal people would look like what people would call as aliens or people from another planet. They looked like us but they were...it was just an image. [What about the crane? Did it look like an ordinary crane?] Yes! It was a normal crawler crane. [It had tracks on it in other words? At that time it was probably about a twenty or thirty tone crane?] I wouldn't know.

[What else did the do there besides that? Did they mess around with your family?] No. They were just putting the canisters in the ground. They just appeared from nowhere. When we got up in the morning they were there...ripping the road out and putting in these giant canisters. [Is it a possibility that it was the government burying some kind of atomic waste from new Mexico that they had hauled down there? Do you know? Hiding it?] There is one problem with that you know because it was an image that they wanted to look like normal people but I could tell that they weren't. [Was it kind of like a hologram?] It was more like a makeup disguise. [Were they wearing black or brown or?] Just everyday cloths. [You know the government always dream up some kind of pocockta suit that they think a normal everyday person would look like and it winds up looking really out of place. Was it anything like that?] No they had...two of them had jumper suits and the rest of them all wore regular cloths...

people. They had…no I don't think it was the government. No it was…for lack of a better word, a race of people that ah…ah…that were doing their own thing, from somewhere else. These canisters they were moving ah…I had never seen that since then. They were round and at lest fifteen feet high and ah...two people tip to tip could go around them.

Since that time they dug up that road and dug up a huge pit. A massive pit! But there was nothing there. Since that time all those houses around there where we were living are gone now. [The company that dug them up was it a regular construction company?] I never seen them before and when we were around there growing up I never seen them again. And ah…I was born there. What were the canister put there for? When they put them in the ground they didn't put them very deep maybe twenty feet. And they had put in steel pieces and was pumping water out like mad. In fact they paid my neighbor to pump water out of the hole. In that area of Florida if you dig very deep you got to pump out the water. Ah…and then you hit lime rock and that goes on for a long ways…I don't know how deep. And that's what's there today a huge lime rock pit and all the houses are gone and there nothing there now. [Why are all the houses gone?] The just bought the property from all the people around there. [Really? Who did it, a big construction company?] I really don't know. [A person could look into that by going through the records at the county I suppose and find out who had bought it?] I believe it would be Levi County, I think.

This is the thing though. When they dug up the holes later on those canisters weren't there and nobody has ever gone up that road and nobody has ever done any construction up there and no one has ever ah...tapered with the ground after they were there the first time. Those canister should have been there because it was like rock... they set it down on a bed of lime rock. I wonder what ever happened to them.

[You said that you and your girlfriend both had some kind of implants put in your neck.?]

No. Ah...I didn't say that. We had two little marks in the back of our necks like a little puncture wound. Just like if you were to take a real fine...like a needle and stick right here in...at the base of the skull and we both had little puncture holes at she same time and neither one of us knew how that happened or how that got there and this was in Louisiana. That was a year ago so that would be ah...March 2002.

[These other guys that came into the room when you were lying on a bed?] No! No, no, no. I was lying on the sofa and I was on my stomach. I had fell asleep when I cam in from offshore. I had taken a flight, I made it in and I laid down to take a nap. I was on my stomach to take a nap facing my kitchen. What it is ah...it's kind of difficult to explain but all of a sudden it was like you were at the dentists office when you were between unconscious and conscious ah...trying to get up but you can't because you were under gas. It felt like that but I wasn't under any kind of gas or nerve agent like that but I did see I felt them and I saw three entities. One of them was a little bigger, one was a little smaller and one was a little smaller. There was three of them and when I woke up...I didn't really wake I felt like they can control things with their minds and suppress your ability to move and function and wake up. I reached out with my left arm. That arm felt like it weighed very, very heavy I could barely move it and I reached out and grabbed one by the shirt like the back of it to the side as they were living. As they were leaving they were trying to get away from me or they were just leaving and I reached out and grabbed one and it was a desperate thing for them to tear loose from me. Finally the other two managed to help him get away from me.

[When was this, a long time ago?] No this was two years ago. [Before you and your girlfriend discovered the puncture marks?] Yea. My girlfriend was at her mother's house. She was at her mother's house down the street at the end of my road. We weren't together when it happened. [Where was that? What town?] Gray! Gray, Louisiana. [Did she talk about any experience like gray aliens?] No, in fact she doesn't remember her dreams. She has never had a dream that she knows of.

[What did the guys look like when you were laying on the couch?] Well, they were human like but they weren't like us. Their heads were different and their eyes were different. Their eyes weren't big almond shaped eyes but they were larger, much larger but they had color. Ah... in their eyes. They weren't black spaces they had color. [Really! You could remember all that! What color? They weren't blue were the?] No. They weren't blue and they weren't orange. Their flesh tones weren't gray like you hear from a lot of people. The clothing was a lot like our T-shirts that we wear but it was a different material.

[When you were hanging onto it what did it feel like?] It felt like ah...the closest thing that I could describe to it would be if you could take silk and blend it with metal and I guess that's the closest I could come to it. It wasn't real thin like a tissue. It was about the thickness of a T-shirt--very strong material. [Like carbon fiber and titanium woven together?] I guess like silk and some type of metal. It was some tough stuff. Nothing like we got. I never grabbed hold of any type of material that felt like that. And you could hold onto it and pull with everything you got and it would not give. It would not stretch. And I had him! He was it and I had him! And it was ah…there were in a panic to get free from me…to the smaller one away from me. And that was something that really happened. I was not in a dream state. It actually happened in my home and right when he tore free, when he got free from my hand it would say it was a period of about eight seconds when I pushed myself up off my sofa and I ran…I ran around my sofa past my aquarium and to the front door. And I ran out to see if I could catch up to them but they had already gone.

People think that things don't exist things don't happen to people but there's animals but extinct for thousands of years but all of a sudden ones pops up you know. And science always proves themselves wrong over and over again. They get a deep revelation or foreknowledge on something they find out what they preached and taught for thirty forty years is all wrong or it wasn't just so it was something else to it. You know. It was the same type of puncture marks and neither one of could figure out how that happened. It happened the same time on the same night and we noticed it the next morning the same time and when she cam down I told her about it I felt it on the back of my neck and she said I got the same thing and I don't know how it happened. {The puncture mark incident was after the incident on the couch. Did they become curious about his tremendous strength and come back to take some brain or DNA samples to find out what kid of humans they were dealing with?}

[did they come back to take some kind of blood samples or DNA after the incident on your couch?] I think so.

[This other thing you mentioned, the glory cloud, what was that?] Hmm…It was in New Orleans. A glory cloud really is and it does exist. Its not a heat lightening ball its just like it says a glory cloud. It is ah…for lack of a better word a golden color. Its' not like misty rain but its thick and its not wet like you were in a cloud. More like vapor. [Did it come in the house?] No the firs time it was at a small church service. [Was it a Baptist church?] No. Nondenominational! It had maybe twenty-five people in there. And a…a guy come up to me and say, "God is trying to tell you something and you need to be anointed and listen." I didn't think nothing of it you know.

[Did it come in the church?] Ah huh. [What size?] Well ah…from that hanging basket to where that wall is right there then it shrunk down to where that distance is where that light is right there. [It was eighteen feet long then it shrunk down to about three feet?] No about nine. And that was a real deal. At that time I had to be about twenty-three. [What color was it?] Well it's a hard thing to describe. Yellow…kind of like the lettering on that hat but it was more like vapor. [Bright gold then?] But ah…It was also like vapor. [Like a golden ball, was it round or a big oblong?] No. It didn't really have a shape other than constantly moving could like is about the closest way to describe it. Lots of people were there and we all were fully alert and aware. [They saw it then?] Oh yea! They saw it, felt it, move in it. They was worshipping. There were about twenty people there.

Also in Gainsville, Florida I saw that ball lightening you were talking about. When we were kids we were in the backyard and I seen it blue, green, red, and white. Ball lightening I seen that. [Was it on the ground or where?] It was everywhere. If you took a Ping-Pong ball and smack it in a small room. It was like that. The white one was small and it would slow down. Come to stop sometimes. It was like at best it would be a half a minute and they were gone again.

[What size where they?] Ah...sometimes they were the size of a basketball and sometimes they were the size of a grapefruit. And ah...sometime they would be three feet across something like that. Sometimes they wouldn't be any bigger than a grapefruit.

Ball lightening can kill you. I saw it come down to this mean guy's porch across the street. It passed right into his chest and he dropped dead as a doornail right there in the spot. He was a real mean guy. If you brought his a flower he'd cuss your ass out you know. He was a mean old goat. It went strait through him and we didn't' see it after it hit him. And that wasn't the first time we saw ball lightening. Right there in Gainsville lots of people seen it. And we were kids you know it would come and go. He fell strait over the porch railing and he was dead before he hit the ground. You just knew he was dead. But ah... it was like you take a piece of paper wad it up through it in the garbage can. I didn't feel any power like you normally feel when someone passes away and the odd thing nobody else felt anything. When you see a little bird fall on the ground and die you feel terrible you start crying but when he died there was nothing there. There was no emotion.

There was seven of us kids that ran around and there was at least five of us that day. We saw a blue, red, and a white one. It was a common thing there for a couple months. A lot of grownup people seen it. My mother for one. She was kind of leery she wouldn't let us go play in the park or nothing. She didn't know what it was. When we go back out it would come around again. Anyway I remember my kid sisters never did see it because they were always in school. It never touched us. It would go around us and in between us like inches you know. You could reach out and try to touch it you but it was way to fast. As quick as you could blink your eyes you never had a chance to touch it. Bank it was gone, bang it was there. Wow there it is again you know. [It seems like the way it moved it was controlled by some kind of intelligence. Do you think ball lightening could be controlled by some kind of intelligence?] Absolutely. I don't know if ball lightening is an intelligent life form or not but there is intelligent life forms that can travel faster than the speed of light...way faster than the speed of light, I believe that absolutely, quicker than the mind can think.

I always thought it was a striking odd thing that none of felt a thing. We grieved over a dead cat or a bird but when this guy dies none of us shed a tear. We could care less. And I don't know what it all was but then he was gone and his house caught on fire immediately afterwards and burnt to the ground. I mean it burnt right to the ground. It was a hellaches fire. In thirty or thirty five minutes that house was completely gone. It was a big house and it just burnt.

Haney cam from across the street and pulled his

body out of the fire. That's right! His name was Haney. Curt...Curt Haney, that was his name that drug that old man out of the fire.

When we were all five years old we would go all over the place you know. We'd go down about a half mile down the street across the railroad tracks. I had a timber wolf for a dog. Thing was a timber wolf but it would travel. He was my dog but he'd travel all over the place you know. I'd hook him up to my tricycle and say lets go lets go and he'd drag me all over the place. My sister would have to hunt for me and it would get dark. I'd be hiding out in the back yard you know and they would be looking all over the place.

I'd go and get bird eggs you know. I thought bird eggs were the prettiest thing. I was infatuated with them. I'd go and collect them by the dozen. I didn't know that I was killing the birds you know.

Anyhow I remember a lot of stuff when I was really little but my sisters don't remember the stuff until I start talking about it but it comes back to em. And there is a dark side.

I could remember once when my sister and I watched...my older sister was in the bed and she would read me a book. And there was a figure of a man in the window. I went to sleep that night and we slept till the next morning. It was in the winter time and we had a big window where my bedroom was at. The person standing there was a dark color. It wasn't a black person. It was a dark color. We both were awake and looked out the window. We tried to move but we could...not...move! You couldn't even turn you head side to side. My eyes were the only thing I could move. I looked to my right hand side and I could see my sister and my sister was looking out the window. It was like something was holding us down and when that figure left from the window we both just sprang strait up. We didn't bend a leg or a knee or our bodies but we both just sprang strait up. Its like you were laying back on the bed and you were launched strait up on your feet. And I couldn't figure out how that happened to this day. My sister doesn't remember but when I start explaining to he she said, "Oh Yea, Wow! I do remember! I can't believe you could remember that." She never brings it up. She doesn't like to talk about the weird things that happen.

[Kevin's sister is now a successful attorney with a photographic memory.]

I remember that when I was little my sister could pick me up with one hand. I remember when I was real small my sister used to hide me in the bread box. I mention this and she said, "You remember that bread box?" "What color was that bread box?" And I'd tell her it was yellow and white the inside of it was tin and she say, "How'd you remember that? You were so small! I thought you had forgot about that." Then I'd tell about the drawer and her hiding me in the bottom of the oven. You know they got that drawer that slides out and then I'd just freak them out when I tell them about the times they wrapped me up in the blanket with my arms down when I was sleeping. I'd wake up and I couldn't move and I'd pee myself and it made me mad you know. And I'd start telling them about this and they start screaming "Oh no! I can't believe you remember that." "You aren't still mad about that are you?" Hell yes I am! And then I'd say pay back's a bitch aint it?

Anyhow I remember stuff way, way back. I

remember there were footprints through our living room and our kitchen. I remember we lived in a wood house had this old style linoleum floor. It looked like a two-legged cow had walked right through our kitchen all around and through our living room and left. Both doors were open but they had been locked. You had better believe it man! Right there in Florida - they have—I don't know what it is about that place. It's just incredible the stuff that happens there.

I have had experience when I was younger with three people. They were like regular people. Ah...but they wore Oriental type clothing and they could think about something and it was there. They could think of a place and they were there in a blink of an eye they could go to another part of the world and they were there. And ah...I was able to levitate with them and they would play games. And they were extreme type people. They were not a peaceful type people. They were a warrior type people. They had a great sense of humor but they had many battles. And a lot of things were going on with them that were like... extreme. But that's the first time that I had any experience in being able to float in the air and move through the air and quickly. [How old were you then?] I had to be at least seven or eight. That was in Galveston, Texas. [You must have moved around a lot.] Yes. At that time the old man did oil and gas exploration.

These three people wore their hair in a pony tails and had gold bracelets and on their wrists and ankles. Their clothing was like oriental silk clothing. [Did they have a dragon pattern on their cloths?] no. Nothing like that. It was more like symbols. They just exuded wisdom, confidence power and you just knew that you were in the presence of absolute power. They could pass through stuff and they could take you on a journey in a blink of an eye. And take you somewhere and in a blink of an eye come right back. I couldn't pass through stuff like they could but they could take you there so quick.

I was explaining stuff about countries I had never visited to my mom and sisters you know, and my sister said, "Kevin, You got the most vivid imagination in the world! I going to check this out." And she did. "What country did you go to." And I'd start explaining to her about the Mongolian flatlands and I'd tell her about the Kurds and the Takagurds, and the Taka this and that and she found out that everything I told her was a fact. I had never seen Mongolia and never been there. So ah…I did that like five or six times with her and finally got pissed off at her and told her to bug off.

I traveled with them bunches of times. I would say that in a span of four of five months I traveled with them a dozen times. They had the capability to fly. They said to me, "You too, you too can fly!" You know they spoke to me in a different language and I understood everything they said. When I spoke to them in English they understood everything I said and when they spoke to me in a different language and I would understand everything they said. And lot of times their voice would sound different. They would talk to me but their lips would be moving but I could still understand everything they'd say. One of them was different. He would explain more. [What would he say?] Oh! He would ask if I wanted to go home or if I wanted to go back and in a blink of an eye I would be back in my back yard or back in my room. You know ah…stuff like that. I was living in Galveston Texas at that time. I had these Harlem Globetrotter pajamas on at the time and I remember different...just feeling different than they were and I remember them saying that it doesn't matter. "Your apparel does not matter."

They were the first to introduce me in the mirror to the fact that the me in the mirror is not the real me. Like a garment when you take off your jacket, it will fall to the floor. They said that one day your flesh like a garment will fall to the floor. They were the first ones to introduce me to this.

I know its hard for people to understand and I know damn well they can't believe it but it's true. It did happen. To me it's not a story. I know its real. It's true it did happen. To me they were a real positive influence. I will remember them forever. I could forget about tomorrow it wouldn't bother me at all for those three characters I will remember them forever. You know what I mean?

I use this term, the physical, the natural order of things, but I think in the supernatural world...in the spiritual the world its so much vaster...bigger than we know about. The earth is like a small grain of sand in a vast eternal dessert. That's how much bigger the spirit world is. If we were to travel we would go to places you can't imagine. You can't comprehend how vaster...larger the other side is. These are my thoughts.

If somebody has information on what kind of canisters were buried near Gainsville, Florida or if you have a story to tell that you want to share with others please write or call. I would be glad to transcribe it for you.

Write:

Henry Kroll
PO Box 526
Kasilof, Alaska 99610

To order copies of the Philosopher's Stone or this book, please go to my web site:

WWW.ALASKAPUBLISHING.COM

About the Author

Henry was born on the family-owned floating cannery February 21, 1944 in the coastal fishing village of Seldovia, Alaska. There he learned to handle a rowboat at age five and at age twenty-two Henry purchased a boat on Chicago and traveled the length of the Mississippi to New Orleans and South via the Inter-coastal Waterway to Mexico. He has written an unpublished book about this voyage title: MISSISSIPPI. Henry studied journalism, physics and chemistry at Sheldon Jackson College, Sitka, Alaska and creative writing, geology and electrical engineering at University of Alaska and Corpus Christi, Texas.

During his life on the sea Henry caught a million salmon, three million pounds of king crab, and accumulating over twenty-five-thousand-days of documented sea time earning his masters' license, all oceans. Henry is the stereotypical Alaskan bush pilot, sea pirate and mad scientist. He attacks his writing and research with the zeal of an apex predator and now has ten books to his credit.

Henry and Mary have three dogs and a plane. They are building a tourist lodge in beautiful Tuxedni Bay, located on the West shore of Cook Inlet in Tuxedni Bay. Henry's scientific research on the inter-dimensional monatomic iridium sands in Tuxedni Bay resulted in his writing, Philosophers Stone. This book contains seven recipes to make a substance in your kitchen that can increase your intelligence and extend your life hundreds of years.

Henry occasionally travels throughout the United States and does guest speaking about political issues, monatomic gold, ancient civilizations, and higher-dimensional spaceships. He will speak at radio stations, UFO conferences, schools and universities. If you have contacts with your local radio station please contact us. Henry will play classical piano for any audience. Henry can be reached by writing: Henry Kroll Tuxedni Bay Lodge 10672 Spur Hwy., #112 Kenai, Alaska 99611.

www.ingramcontent.com/pod-product-compliance
Lightning Source LLC
Chambersburg PA
CBHW032023170526
45157CB00002B/832